P9-DVD-944

PENGUIN BOOKS

BUZZ

Stephen Braun is an award-winning science
writer and television producer living in
Boston. He is currently Executive Producer
at the New England Research Institutes.

buzz

THE SCIENCE AND LORE
OF ALCOHOL AND CAFFEINE

Stephen Braun

PENGUIN BOOKS

PENGUIN BOOKS
Published by the Penguin Group
Penguin Putnam Inc., 375 Hudson Street,
New York, New York 10014, U.S.A.
Penguin Books Ltd, 27 Wrights Lane, London W8 5TZ, England
Penguin Books Australia Ltd, Ringwood, Victoria, Australia
Penguin Books Canada Ltd, 10 Alcorn Avenue,
Toronto, Ontario, Canada M4V 3B2
Penguin Books (N.Z.) Ltd, 182–190 Wairau Road,
Auckland 10, New Zealand

Penguin Books Ltd, Registered Offices:
Harmondsworth, Middlesex, England

First published in the United States of America by Oxford University Press, Inc. 1996
Published in Penguin Books 1997

10 9 8 7 6 5 4 3 2 1

Copyright © Stephen Braun, 1996
All rights reserved

The author and publisher thank the following for
permission to reprint specified material:

From *One Fish, Two Fish, Red Fish, Blue Fish* by Dr. Seuss. TM
and copyright © 1960 and renewed 1988 by Dr. Seuss Enterprises,
L.P. Reprinted by permission of Random House, Inc.

Excerpted with permission of Scribner, a division of Simon & Schuster,
from *A Moveable Feast* by Ernest Hemingway. Copyright © 1964 by Mary Hemingway.
Copyright renewed 1992 by John H. Hemingway, Patrick Hemingway,
and Gregory Hemingway.

THE LIBRARY OF CONGRESS HAS CATALOGUED THE HARDCOVER AS FOLLOWS:
Braun, Stephen.
Buzz: the science and lore of alcohol and caffeine / by Stephen Braun
p.   cm.
Includes bibliographical references and index.
ISBN 0-19-509289-9 (hc.)
ISBN 0 14 02.6845 6 (pbk.)
1. Alcohol—Popular works.   2. Caffeine—Popular works.
I. Title.
QP801.A3B73   1996
615´.7828—dc20   95–47790

Printed in the United States of America
Set in Transitional

Except in the United States of America, this book is sold subject to the condition
that it shall not, by way of trade or otherwise, be lent, re-sold, hired out, or otherwise
circulated without the publisher's prior consent in any form of binding or cover other
than that in which it is published and without a similar condition including this
condition being imposed on the subsequent purchaser.

To the memory of

Robert Arnold Braun

Shaman, Scientist, Father

Acknowledgments

The idea for this book germinated during a fellowship in neurobiology at the Marine Biological Laboratory in Woods Hole, Massachusetts. There I was introduced to a radical new understanding of the brain and the way it works—and there, too, I saw clearly for the first time how substances such as alcohol and caffeine could affect the machinery of the mind. I am indebted to Irwin Levitan, the scientist in charge of the lab in which I was a student, for his patient explanations, steadfast encouragement, and infectious enthusiasm for neuroscience. To the MBL I owe thanks as well, for offering science journalists such a unique opportunity to experience science by *doing* it, rather than simply reporting it.

Buzz has benefited tremendously from the keen-eyed review of many scientists. Foremost among them is Steven N. Treistman, who carefully and thoughtfully read the entire manuscript and who was an unflagging supporter of the project from day one. In addition, the following scientists took time from busy schedules to read selected chapters or chapter portions: Gary Kaplan, Thomas Dunwiddie, Barry Green, Robert Greene, and Annette Rossingnol. Other scientists helped by answering often lengthy lists of questions and by providing copies of papers on relevant subjects. My thanks to Robert D. Blitzer, Joseph Brand, James Brundage, Michael Charness, John Daly, David Lovinger, Quentin Regestein, Forrest Weight, and Mark Whitehead.

For the excellent molecular models included in the book, I thank Joe Gambino of the University of Massachusetts Medical Center. The superb line drawings of ion channels and the neuronal synapse are the work of Ann Bliss Pilcher.

I am grateful to Howie Frazin and Lynn Prowitt for deftly finding the confusing parts and helping me clarify them. My editor at Oxford University Press, Kirk Jensen, provided invaluable guidance and repeatedly led me to the correct voice for the book. I am also most grateful to copy editor Gail Weiss, whose many valuable suggestions improved *Buzz* considerably.

Finally, my heartfelt thanks to Mary Anna Towler for teaching me to write; to Steve Pilcher, Tim Braun, and Doug Beyers for invaluable lessons in perceptual relativity; to Oralee Stiles for the fabulous quote; and, above all, to my wife, Susan Redditt, for patience far beyond the call of duty. Were it not for her loving support in matters great and small, this book would not have been possible.

Contents

buzz

The government of a nation is often decided over
cup of coffee, or the fate of empires changed
an extra bottle of Johannisberg.

—Duc de Richelieu

Introduction

Aristotle, in *The Problemata*, posed the following questions: Why are the drunken more easily moved to tears? Why is it that to those who are very drunk everything seems to revolve in a circle? Why is it that those who are drunk are incapable of having sexual intercourse? Aristotle considered the brain nothing more than a radiator for cooling blood, so it's not surpising that he couldn't answer these questions. He and others attributed the intoxicating powers of wine and beer to mysterious "spirits" of inebriation.

In a similar vein, seventeenth-century doctors puzzled over

the stimulating effects of coffee and tea. Some argued that the beverages contained "cold and moist" essences that altered the balance of the body's four vital fluids, or humors. Others, however, thought that coffee and tea should be classified as "warm and dry" in the humoral spectrum. The issue was never resolved, though it exercised some of the era's best medical minds for decades.

Alcohol and caffeiene today are the world's most widely consumed mind altering substances, and people are as curious as ever about what they are and how they work. Like Aristotle, they wonder why, when they're drunk, they see things spinning and why alcohol can deaden sexual response. In addition, people have new questions, arising as often from media reports of scientific studies as from popular myth. Do women become intoxicated more easily than men? Does caffeine worsen premenstrual symptoms? Is alcoholism a genetic disease? Is caffeine bad for you or isn't it? Does alcohol really kill brain cells? Can caffeine help you lose weight?

The answers to such questions elude even sophisticated consumers—those who know their cabernet sauvignon from their sauvignon blanc, and their Kenya AA from their Aged Sumatra. The reason is simple: until very recently, *nobody* has been able to answer these questions. It's not that alcohol and caffeine are terribly complex or difficult to understand. In fact, they are rather simple molecules, the structure of which has long been known. The problem is that the *target* of those molecules, the human brain, *is* terribly complex and difficult to understand. Progress in learning how alcohol and caffeine work has had to wait for new knowledge of how the brain works.

Fortunately, in the past two decades the science of the brain—neuroscience—has blossomed. New investigative techniques have opened up the black box of the brain and have begun to shed light on its inner workings. Out of this new insight has come a radically improved understanding of how alcohol and caffeine work. Much of this information is so new that it is known only to certain research scientists and people who read scientific journals. And on the rare occasion that new findings *have* made it from the lab bench to the corner bar, the message often arrives garbled.

For example, an early study suggesting that alcoholics seem to have fewer brain cells than nonalcoholics was widely publicized and contributed to the idea that alcohol kills brain cells (Harper and Krill 1990). Later studies clearly disproved this idea, but the notion remains. Another common idea, based on a scientific paper investigating the metabolic effects of caffeine, is that drinking coffee helps burn off fat (Costill et al. 1978). There is a grain of truth to this idea, but as we'll see, it's a very small grain indeed.

Advances in neuroscience have also turned some long-standing ideas on their head. For instance, many people learn in high school that alcohol is a depressant—a kind of chemical sledgehammer for the mind. Alcohol is actually much more complex. It produces effects that mimic those of many other drugs, such as opium, cocaine, Valium®, and ether. The result is an intoxication far more dynamic and complicated than most people realize. Caffeine is similarly misunderstood. It is not a direct stimulant, like a shot of adrenaline. Instead, it works indirectly by interfering with one of the brain's main chemical "brakes." Like a car with a sticky brake pedal, the brain speeds up because it can't slow down.

Some of the new findings, though, support previously unsubstantiated folk wisdom about alcohol and caffeine. Moderate doses of alcohol, for instance, have been shown to have some health benefits, such as reducing the risk of cardiovascular disease. And, as many athletes have long suspected, caffeine has been found to significantly boost performance in a number of sports, such as running and bicycling.

In short, recent scientific progress has revolutionized the understanding of alcohol and caffeine. This revolution has not yet been recognized by the public, despite the fact that these two substances are consumed daily by a majority of human beings on the planet. Caffeine is the most popular drug on earth (Gilbert 1984). It is contained in tea, coffee, cocoa, chocolate products, many soft drinks, and more than 2,000 nonprescription drugs (Robson 1992). In the United States, roughly 80 percent of adults consume caffeine in one form or another *every day* (Strain et al. 1994).

In the United States, alcohol is second only to caffeine as the drug of choice. Sixty-eight percent of American men and about 47 percent of American woman say they drink alcoholic beverages at least occasionally (Department of Health and Human Services 1993). In other countries where the rate of smoking is relatively high, nicotine takes second place, nudging alcohol to the number-three position. Thus, with the exception of water, *all* of the most popular beverages on earth contain either caffeine or alcohol.

This book is about how these two phenomenally popular substances work: how they affect the brain and the body, and how they are currently understood in light of recent scientific advances. This information is laced throughout with cultural history and personal stories. Mikhail Gorbachev, David Letterman, F. Scott Fitzgerald, William Shakespeare, Buddhist monks, and Arabian goatherds make cameo appearances. We'll hear what Johann Sebastian Bach, Teddy Roosevelt, and John Steinbeck have to say about alcohol and caffeine. We'll meet a champion swimmer suspended from competition for two years because she ingested too much caffeine, and a group of college students who helped reveal how alcohol affects sexual response.

Human stories such as these are inevitable because alcohol and caffeine are so deeply and inextricably woven into the fabric of daily life for most of the earth's population. They are familiar drugs—as close and comfortable as a cup of coffee or a can of beer. And yet they and the buzz they produce remain for most people just as mysterious and unpredictable as the spirits they were once thought to be.

Wine is as good as life to a man, if it be drunk moderately: what life is then to a man that is without wine? For it was made to make men glad.

—*Ecclesiasticus*

Alcohol 101

The Green Dragon

In the mid-1980s, citizens of the Soviet Union were faced with a nationwide shortage of sugar so serious that this basic food-stuff was rationed like gasoline. The cause of the shortage lay not in the usual culprits of inefficient state-run production, inadequate imports, or dilapidated distribution systems. No, it was *zelyony zmei*—the green dragon. That was the nickname given to Mikhail Gorbachev's crusade to wean his countrymen from their national drink: vodka. The name referred to the coils of often greenish copper pipes used in the thousands of home distilleries built to supply what the government was trying to limit.

The nation's sugar supply was disappearing into these secret, jury-rigged stills. Of course, Soviet moonshiners also made booze from potatoes, corn, wheat, barley, and other starchy materials. But these raw ingredients complicated the process. To be useful for fermentation, starch molecules first have to be broken down by enzymes into the individual sugar molecules from which they are made. It is much simpler and neater—though not necessarily cheaper—simply to start with raw sugar. Soviet bootleggers pouring bags of sugar into their home fermentation vats doubtless do not dwell on the details of the transformation they shepherd: the conversion of ordinary table sugar into a powerful mind-altering drug. But it is a remarkable feat of alchemy indeed.

For most of human history, this transformation was a deep mystery, even though beer- and wine-making are among civilization's earliest professions. Recent analysis of a yellowish substance found in the bottom of clay jars unearthed in Iran confirms that the Sumerians were accomplished brewers 5,500 years ago (Wilford 1992). And brewing wasn't a trivial enterprise. One of the most common pictographs in Sumerian ruins is the sign for beer—hatch marks that represent the crisscross patterns dug into the bottom of beer vessels.

The Sumerians are the most ancient users of alcohol that we know about. But the discovery of fermentation was a worldwide phenomenon. Where there was sugar, humans learned ways to encourage its conversion into alcohol by fermentation. But although the art of fermentation is ancient, the science is not. As civilization flourished over the centuries, people made beer and wine with only the crudest understanding of what they were doing. It wasn't until quite late in the game that anyone had a clue about what alcohol was or how it was produced. Understanding these two things is the first step in understanding such larger issues as intoxication, hangovers, and addiction—topics we'll explore in later chapters.

Here we'll simply get to know alcohol for what it really is.

Portrait

If you could examine beer, wine, or liquors under a super-powerful microscope, you would see a wild jumble of molecules bashing into one another in a confusing tumult. If you zeroed in on a single molecule of alcohol, here's what you would see:

Ethanol

This is what people get when they order a drink at a bar—a type of alcohol called ethanol. You can see that its nine constituent atoms nestle against one another, forming a lumpy particle that, in this particular orientation, bears more than a passing resemblance to an exceptionally pudgy dog.

Ethanol is made up of two carbon atoms (shown in black), an oxygen atom (in gray), and six hydrogen atoms (in white). Notice the "head" of the dog. It's the oxygen–hydrogen group on the upper right. Any molecule with this group attached to a carbon atom is called an alcohol. Since this is a fairly common arrangement in molecules, there are *lots* of alcohols in the world. Glycol is the alcohol in antifreeze, and cholesterol is a complicated type of alcohol vital for many bodily functions (and in excess, helps to clog blood vessels). Ethanol is simply the most famous member of a very large family of alcohols, though it's not the only alcohol that causes inebriation. You also can get drunk on ethanol's simpler cousin, methanol, which can be produced by the fermentation of wood and thus has the common name wood alcohol:

Methanol

Although cheap to produce, methanol has a rather serious drawback as an intoxicant: it is broken down in the body by an enzyme found in particular abundance in the retinas of the eyes. This enzyme converts methanol into a closely related molecule: formaldehyde. As anyone who has dissected a preserved frog knows, formaldehyde is *not* something you particularly want inside your eyeballs. But that is exactly what happens when people drink methanol. The enzyme in their retinas converts methanol to formaldehyde, causing permanent blindness. Ethanol's extra carbon atom gives it a shape just different enough from that of methanol to be unaffected by that enzyme in our retinas. We thus can drink freely with the knowledge that although we may not be clearheaded the next morning, we will at least be clear-eyed enough to find the medicine cabinet.

Another of ethanol's interesting characteristics is its diminutive stature. Compare ethanol with some other molecules that produce intriguing results when introduced into human brains:

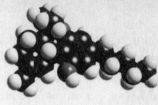

Ethanol   Morphine   Tetrahydrocannabinol (THC)

Morphine and THC are obviously larger and more complex than ethanol. But even these drugs are tiny compared with the molecules common in the human body. Here, again drawn to scale, is ethanol and hemoglobin, the molecule that carries oxygen in red blood cells:

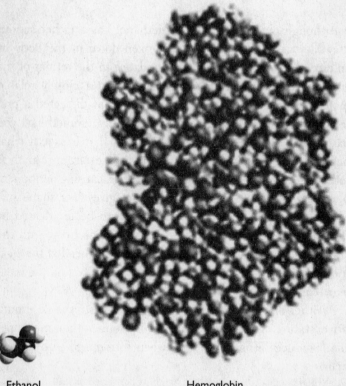

Ethanol   Hemoglobin

Most of the molecules affected by alcohol—including those involved with intoxication—belong to this jumbo-size class of molecules.

Of course, ethanol isn't just *relatively* small. It's important to appreciate just how many ethanol molecules are found in a standard drink—that is, a 12-ounce can of beer, a 5-ounce glass of wine, or a 1.5-ounce shot of liquor. Each of these drinks typically contains a half-ounce of pure ethanol, or approximately 200 *quintillion* ethanol molecules, each of which has the ability to disrupt some part of your body's cellular machinery.

Size isn't the only interesting aspect of ethanol's structure. Ethanol molecules also carry a small electric charge that is crucial to its behavior in the body. The oxygen atom in the "head" of the molecule makes that region slightly negative. This charge meshes nicely with the slight positive charge on one side of water molecules. It looks something like this:

Ethanol   Water

Without the electric charge of its "head" region, ethanol would separate from water like oil. Not only would this make mixing a gin and tonic an exercise in frustration, but it would make it difficult for ethanol to penetrate our water-filled bodies. Oddly enough, however, ethanol is *also* soluble in oils and fats. That's because the "tail" end of the molecule is *not* significantly charged. This region would prefer, electrically speaking, not to associate with water, and so it meshes easily with fat and oil molecules that would also rather avoid water. Ethanol is thus a powerful solvent that can roam freely throughout the body. Because it dissolves easily in water, it is rapidly absorbed from the digestive tract and mixes easily with blood. Because it also dissolves in fats, it freely passes through cell membranes, which are basically double-walled bubbles of fat.

The overall structural shape of ethanol is important as well. A key idea in chemistry is that the physical *shape* of a molecule can be critical to its behavior in the world. The knobby protrusions and infolding pockets of most molecules keep them from touching in a significant manner. The bumps usually get in the way. But sometimes one molecule's knob will happen to fit just perfectly into another molecule's pocket like a key in a lock. When that occurs, all sorts of interesting things can happen. The two molecules may warp or twist slightly under the influence of the connection. Normally stable molecules may suddenly become unstable or split altogether.

As we'll see, these physical shape-to-shape interactions underlie many of ethanol's effects in our brains and bodies. The way that ethanol's shape matches the particular shape of certain key proteins and other molecules in the brain has everything to do with why we get intoxicated.

Yeast Trash

Now we know that the "spirits" in wine, beer, and liquor are really quadrillions of ethanol molecules. But exactly where do those molecules come from? Considering that humans have been making alcohol for at least 5,500 years, it has taken a surprisingly long time to figure this out. Understanding alcohol's origins will shed light on some minor puzzles about how it behaves in the body.

Soviet bootleggers know part of the answer. Alcohol comes from sugar. They also know that this transformation isn't spontaneous—it requires the assistance of one-celled fungi called yeast. This much has been known for centuries. But it wasn't until 1939 that the full details of the story were finally worked out by a scientist named Embden O. Meyerhof. Meyerhof was trying to figure out how sugar is metabolized in our bodies; when he solved this problem, he also had the answer to the age-old question of where alcohol comes from.

The process starts with glucose, which is the sugar both humans and yeast use to power their bodies. Other sugars, such as the sucrose (table sugar) used by the Soviet home distillers, must be converted to glucose before things get going, chemically speaking. Like humans, yeast cells prefer to burn their glucose with oxygen to produce energy. But yeast cells sometimes find themselves in situations where oxygen is scarce— for instance, when they are trapped in the bottom of huge vats of grape juice. In such circumstances, they manage to carry on by using backup metabolic machinery designed to burn glucose *without* oxygen. This anaerobic system is much less efficient than the primary, oxygen-using system. Instead of gradually (and completely) breaking down the glucose molecule with oxygen to release lots of energy, the anaerobic system simply splits the glucose in two, which results in a relatively feeble amount of energy—just enough to sustain minimum life functions until oxygen returns.

This splitting isn't accomplished in a single whack, like a log split with an ax. One of Meyerhof's contributions was the discovery that it takes ten separate steps to split glucose without oxygen. A series of ten separate enzymes is used to twist and pick apart the original glucose molecule until it can easily be cleaved. The process resembles nothing as much as a molecular disassembly line in which the glucose molecule is worked on by one enzyme and then is passed on to the next and so on. The details of that process are interesting in their own right, but all we're really concerned with here are those two shards remaining after the glucose is finally split. Those shards are molecules of ethanol.

The birth of alcohol via this inefficient splitting of glucose has one very salient consequence for humans: most of the chemical energy of the original glucose molecule remains bound up in the ethanol fragments. That energy equals calories: about seven per gram—which works out to about a hundred calories in a standard drink from the alcohol alone. Alcohol, in other words, is no diet drink.

Alcohol's origins also explain some facts about the alcohol content of some common drinks. Yeast cells struggling to survive under suffocating conditions quickly excrete the ethanol fragments because they are basically poisonous. Ethanol interferes with many of the reactions vital to the life of a cell. As a result, yeasts excrete ethanol, which slowly builds up in the surrounding liquid—exactly where the brewer or vintner wants it. Given an adequate amount of glucose, the ethanol content of a fermenting liquid rises until it reaches about 12 percent. At this point, it starts to back up inside the yeast cells because it can no longer diffuse across the cell wall. Unable to dispose of the poisonous waste, the yeasts shut down and become dormant. All activity stops, including the production of new ethanol. This is the reason that most table wines have roughly a 12 percent alcohol content: that's as high as it can go before the yeasts throw in the towel. Some wines can achieve slightly higher values if they are unusually rich in glucose, but the only way to get significantly higher ethanol levels is by distillation— the gentle boiling of a liquid in a sealed container.

Distillation, by the way, is possible only because ethanol molecules happen to evaporate more quickly than water molecules from a liquid mixture such as wine or beer. Water molecules carry an electric charge and tend to stick to one another.

Ethanol, however, is only weakly charged—and only at one end. That leaves ethanol molecules free to escape at temperatures lower than the boiling point of water. If these free ethanol molecules are captured (by condensation coils, for instance), any desired concentration up to 96 percent can be achieved. A 100 percent ethanol solution is impossible to achieve with ordinary distillation techniques because some water molecules unavoidably escape along with the ethanol. Commercially pure ethanol is produced using a variety of chemical reactions to eliminate the water.

Of course, the ethanol concentration of beverages is measured in two ways: by percent alcohol and by "proof." The term "proof" dates to the seventeenth century when various means were devised for checking, or "proving," that a beverage had the alcohol content its label claimed. The proof number is just about double the percentage by volume of alcohol. A wine with 12 percent alcohol, in other words, is roughly 24 proof.

The Human Connection

We've now seen that ethanol is molecular garbage made by yeasts when they burn glucose under the duress of suffocation. As remote as this situation is from daily life, there is, in fact, an interesting connection between yeast and humans. It turns out that we have almost exactly the same cellular machinery in our bodies that yeasts do—and for the same reason.

Like yeast, the cells in our bodies usually burn glucose with oxygen because it releases so much energy. But even for highly mobile humans, oxygen isn't always available. In fact, it's exactly those parts of the body that are used most vigorously that may face an oxygen shortfall. In strenuous running, for example, certain muscles use oxygen more quickly than it can be replenished by the blood, leading to a localized condition reminiscent of that faced by yeast in the bottom of fermentation vats. At such times, muscle cells fall back on the same ineffi-cient anaerobic machinery used by yeast—machinery we in-herited from our single-celled ancestors.

Actually, the entire molecular apparatus is retained except for a small, but important, alteration. Since yeasts are tiny or-ganisms, they can easily dispose of their ethanol trash. Hu-mans, however, being relatively enormous, must rely on a complicated waste-elimination system. If we produced ethanol as a waste product of metabolism, as do yeast cells, it would circulate in our blood and we'd end up rip-roaring drunk when-ever we exercised hard. This obviously wouldn't be a very adap-tive situation from an evolutionary standpoint. It's not surprising, therefore, that in humans and other animals the process of anaerobic metabolism is slightly altered to avoid the production of ethanol. In humans, the last of the ten steps required to split glucose is changed so that the two resulting fragments are molecules of lactic acid, not ethanol. Too much lactic acid, of course, can cause muscle soreness and fatigue, but at least it left our forebears sober while they fled from saber-toothed tigers and their ilk.

Molecular Spirits

This chapter has introduced a new way of looking at alcoholic beverages. The "spirits" in these drinks are actually trillions upon trillions of individual ethanol molecules. These molecules are small and knobby and have some peculiar attributes, such as the ability to dissolve in both water and fats—characteristics that are key to their ability to breach the body's normal de-fenses. We've also seen that alcohol is actually a type of ex-crement produced by suffocating yeast. It exists as a metabolic waste product and would be produced by humans as well except that nature apparently didn't select for creatures with this ability.

All of this adds up to a rather uncommon perspective on alcohol. But it is not yet a particularly helpful perspective because as interesting as it may be, this new understanding of alcohol's true nature is only half the equation of intoxication. The other half is the human brain.

It smells like gangrene starting in a mildewed silo, it tastes like the wrath to come, and when you absorb a deep swig of it you have all the sensations of having swallowed a lighted kerosene lamp.

—Irwin S. Cobb, on corn whiskey

Down the Hatch

2

A Wee Dram

Imagine that you find yourself floating, mysteriously, in a shot glass containing a dram of a nice scotch whisky—let's say it's eighteen-year-old Macallan. You've been shrunk (along with appropriate scuba gear and a powerful flashlight) by a factor of about a billion to the size of a small molecule. At this scale, a single ethanol molecule is roughly the size of a corpulent Labrador retriever.

As you glance around, you see ethanol molecules wiggling in all directions, along with similarly frenetic water molecules and, here and there, larger, more globular molecules you can't identify. These are probably the phenols, sugars, tannins, and in-organic compounds that give the scotch its amber color, smoky aroma, and distinctive taste. Many of these molecules seeped into the initially crystal-clear whisky from the oak casks in which the liquor was aged. This particular whisky is aged in oak barrels used previously for sherry. That means that some of the molecules originated in the grapes that went into the sherry. Some of the other molecules you notice originated in the smoke from the peat fires used to roast the barley malt prior to fermentation, and still others are derived from the barley grains themselves.

All in all, the view from inside this shot glass is remarkable. But you have no time to appreciate it. Suddenly you are caught in a tidelike surge. The glass is raised ceremoniously (we'll assume in a toast to science rather than Bacchus) and then tipped into some anonymous mouth. It is suddenly dark.

The imaginary voyage you're about to take will begin our exploration of how the ethanol we met in the previous chapter actually works in the human body. In this chapter, we'll see what happens as ethanol moves from the mouth to the stomach, into the blood, and then to the liver, an organ with a key role in the experience of intoxication. As you take this trip, you'll learn about a number of the smaller mysteries of drinking, such as why many people prefer drinks "on the rocks" and why some people are much more sensitive to alcohol than others. You'll also be introduced to some of the fundamental ideas that are key to understanding how both alcohol and caffeine work in the different parts of the body.

Tip of the Tongue

You're now flowing in a river of molecules across the furrowed surface of an enormous writhing slab of muscle: the tongue. You flick on your light. Although most of the flow is surging toward the back of the throat, you and several hundred other molecules are being driven by the currents down into a deep canyon.

As you descend, a mushroom-shaped structure looms up out of the murk. At your molecular scale, it appears monstrous. You're looking at a fungiform papilla—one of the 9,000 or so small bumps on the tongue that most people call taste buds. The name "fungiform" refers to the mushroom shape of these bumps, which are actually just support structures. Each papilla contains between fifty and one hundred individual taste buds, most of them located on the *underside* of the papilla, not the top.

The current forces you and your company of ethanol molecules up against the papilla. Dead ahead, a taste bud comes into view. Not a bud at all, it looks more like a gaping hole in the surface of the papilla. Down into the hole you plunge. At the bottom of the pit, you see what looks like a bed of kelp waving slowly in what is now a rather viscous mixture of scotch and saliva. These are microvilli—the fingerlike projections of the many individual taste receptor cells in each bud. The fingers greatly increase the surface area exposed to the molecule-laden currents washing in and out of the taste bud's pore.

As you close in on the microvilli, they begin to tower over you like giant pillars. Now you're so close you can see tiny bumps studding the otherwise relatively smooth surface of the microvilli fingers. A sudden wave shoves you against the pillar near one of the bumps. You can now see the bump for what it really is: a single, huge protein molecule, lodged in the membrane of the microvillus like a tennis ball stuck in a chain-link fence. You're looking at the part of the "ball" exposed to the outside world, but the molecule spans the cell membrane and extends back into the body of the microvillus itself.

Suddenly the bump moves spasmodically. The entire molecule rapidly changes shape. One moment it looks like a solid, protruding mass from the cell membrane, and the next moment, with a quick twisting motion, it opens up, forming what looks like a hole or a tunnel. Aiming your light down into it, you can see right through the membrane to the cell interior. From what you can see, the interior is densely packed with a bewildering array of molecules in all shapes and sizes. With the tunnel open, you notice thousands of tiny particles surging through the breach. These are actually electrically charged atoms—generically called ions—and their surging is one of the fundamental actions underlying both normal thought and the altered states induced by alcohol and caffeine.

The quivering, dynamic molecule you've been watching is called an ion channel, because it acts like a gate regulating the flow of ions across a cell membrane. Since ions carry electric charge, the flow changes the electrical environment inside a cell. Alcohol and caffeine can affect the way ion channels open and close (Figures 1 and 2). Like a foot in the door, they can leave a channel stuck open. Or, alternatively, they can make it harder for a channel to open in the first place. It all depends on the type of channel involved. In later chapters, we'll look at ion channels in the brain as well as other parts of the body and see what happens when they are influenced by alcohol or caffeine (or both).

Here on the tongue, ion channels play a critical role in our sense of taste. The channel you've been watching is embedded in the membrane of a taste receptor cell. The part of the channel that protrudes into the scotch-laden currents contains certain spots that "recognize" the shapes of some of the larger molecules floating around you. If one of these molecules sloshes against one of these spots on the receptor, it binds very briefly, causing the entire physical assembly to shift into the open configuration. Then, in a blink of an eye, the trigger mol-

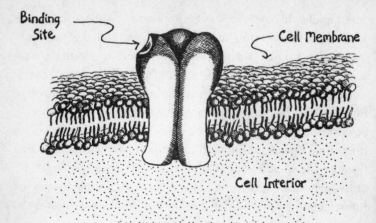

Figure 1. Closed ion channel, cross section. (Ann Bliss Pilcher)

ecule detaches and the ion channel resumes its normal closed shape. If enough ion channels open at approximately the same time (as is happening now from the sudden influx of scotch) the electrical balance of the entire taste receptor is tipped so much that nearby nerves are stimulated. In this case, those nerves fire off a message that is interpreted in the brain as mildly bitter. Perhaps the molecules we've been looking at, then, are sharp-tasting tannins derived, originally, from the oak sheaves of the sherry casks.

This same sequence of events is happening all over the tongue. The receptors for the four basic tastes—salt, sour, sweet, and bitter—are all stimulated to one degree or another by the molecules in the scotch. The end result is a complicated taste message sent to the brain—a message made even more complex by the simultaneous reception of signals from the nose, which has a much more diverse set of receptors than the tongue.

Interestingly, none of these taste signals is due to ethanol. Pure ethanol is virtually tasteless, though some people report that in a weak solution, ethanol tastes faintly sweet. But ethanol clearly *does* interact with a separate set of nerve receptors in our mouth, nose, and esophagus called polymodal pain fibers. These receptors outnumber taste receptors by about two to one. They are very sophisticated nerve cells that respond to three kinds of stimuli: physical pressure, temperature, and specific chemicals. When these receptors are overstimulated, we perceive pain or irritation.

From your current vantage point, you are well positioned to see how alcohol can get at these pain fibers. Drifting away from the quivering ion channel, you notice legions of ethanol mol-

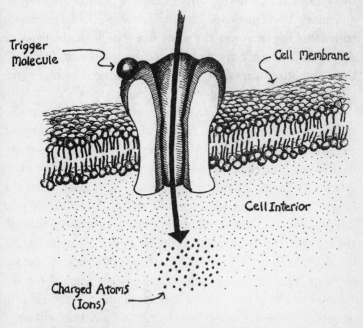

Figure 2. Open ion channel, cross section. (Ann Bliss Pilcher)

ecules disappearing across the membranes of nearby cells. Because they're made of fat, membranes repel most of the molecules wandering around outside a cell, including the ubiquitous water molecules. But this normally impenetrable barrier is easily breached by fat-soluble ethanol molecules, which slip through like little ghosts. Since you are *not* fat soluble, you can't follow the ethanol into nearby cells. But if you could, you'd see them passing rapidly through several layers of cells to the polymodal pain fibers lying underneath.

Exactly how ethanol stimulates these fibers is a mystery at the moment. But anyone who has tasted undiluted whiskey, gin, rum, vodka, or any other high-proof drink can attest that they are, indeed, stimulated by ethanol. Research has shown that the more these fibers are stimulated by an alcoholic drink, the greater the burning and irritation in the mouth, throat, and nose. We experience burning because ethanol somehow stimulates the receptors the same way that high temperature does. (Ethanol isn't the only molecule capable of "tricking" these receptors. Capsaicin, a molecule produced by many species of pepper plants, does the same thing.)

As most drinkers know from experience, these burning sensations can be reduced significantly by chilling the liquid prior to consumption. Chilling actually serves several purposes. First, cool ethanol molecules have less vibrational energy than warm ethanol molecules. Less energy means less impact when the molecules physically bump into the mouth and throat's pain receptors. It also reduces the ability of ethanol to move through the layers of skin cells to get to the receptors in the first place. Cooling also makes it harder for ethanol molecules to escape as vapor. This effect is particularly important when the liquid is contained in glasses such as brandy snifters. If the concentration of trapped ethanol vapor is high, a quick sniff will prove

uncomfortable—if not downright painful—which may interfere with one's appreciation of the subtler aromas of the drink.
Finally, a cold drink *directly* stimulates receptors for coolness.
When these receptors are stimulated, their signals appear to
partially offset—or at least muddle—the simultaneous signals
for heat generated by the false stimulation of the nearby polymodal pain fibers.

It's for all these reasons that some people prefer ice-cold
beer, chilled wine, and liquor on the rocks. Others drink, at
least in part, to savor the flavor and aroma of a drink—the
complex butterscotch and vanilla flavors of aged bourbon, the
raspberry and cherry notes of a good Zinfandel, or the nutty
sweet flavors of a good tawny port, for instance. For these folks,
temperature control is a delicate balance. Chill a drink too
much, and the aromas and flavors are lost. But let it get too
warm, and excessive ethanol is released, interfering with some
of the more pleasant aspects of the drink.

Front-Line Defense

By now, as you lie snagged in the labyrinthine corridors of the
tongue, most of the scotch is long gone. The bulk of the shot
passed immediately through the mouth and down the esophagus, where, along the way, ethanol molecules stimulated polymodal pain fibers and generated a gentle burning sensation
as it dropped into the stomach. Now, with a swish and a swal-
low, you, too, are sent on your way.

As you fall into this cavernous organ and look around, you
see that the shot was administered before a meal. The stomach
is mostly empty—the whisky is lying in a shallow pool where
it is now mixed with highly acidic gastric juices. The acid levels
here are 100,000 times higher than those found in the blood—

high enough to destroy most of the larger molecules in the whisky. But ethanol molecules, because they are so small and stable, are immune to acidic destruction.

As you near the corrugated wall of the stomach, your light glistens off the thick layer of mucus that protects the stomach from its own acid. But even this normally impervious mucus is easily breached by ethanol. As you watch ethanol molecules disappearing into the mucus, you notice an increased oozing from many tiny pits in the lining itself. Tiny glands at the bottom of these pits produce stomach acid. Ethanol stimulates these glands to produce more acid by means that are not yet understood. The increased acid goes unnoticed by most people, but those with either sensitive stomachs or large appetites for alcohol may experience stomach pain or indigestion from the excess acid.

As they pass through the stomach wall, ethanol molecules also stimulate receptor cells sensitive to such variables as acidity, distention, and the presence of poisons such as bacterial toxins. If enough of these so-called sentry receptors are stimulated by ethanol, they send an electrical signal up to a small portion of the brain called the emetic center. A less polite name would be the "vomit center" because this cluster of neurons controls the involuntary muscle movements of retching. Fortunately, these sentry receptors are not terribly sensitive to ethanol: most people have to drink quite a bit to trigger them. But if these cells are already responding to stomach distention, such as from a large meal or one-too-many beers, then the added stimulation from ethanol can increase the chances for an unwelcome return of the evening's ingestion.

Since you are so close to the stomach lining, you decide to take a swim through the mucus and follow some ethanol molecules. Contrary to previous beliefs, the stomach is *not* the primary route by which ethanol enters the blood. Some absorption does take place, but it is generally insignificant to the experience of intoxication. One reason for this becomes clear as you slip and slither between mucous molecules. Directly ahead you see a huge, globular molecule the size of a two-car garage. It's part of the body's first line of defense against ethanol: a detoxifying enzyme called alcohol dehydrogenase. You watch as an ethanol molecule bumps up against the enzyme. Nothing happens. The ethanol rebounds, twists, and suddenly wedges tight into a small crevice in the face of the enzyme. Instantly, one of the hydrogen atoms on the ethanol molecule is ripped off. Shorn of its hydrogen atom (hence, dehydrogenated), the ethanol is released by the enzyme, which is now ready to take on the next ethanol molecule that happens its way.

The molecule left behind by this surgery is no longer ethanol. It is called acetaldehyde. The removal of that single hydrogen atom renders ethanol pharamacologically inactive: you can't get drunk on acetaldehyde. Nonetheless, acetaldehyde is a very chemically reactive little molecule that, like ethanol, can seriously interfere with the molecular machinery of a healthy cell. Specifically, it readily binds with a wide range of proteins, which can cripple their normal functioning. That's why your body is equipped with a *second* enzyme in its defense against alcohol; called aldehyde dehydrogenase, it is specifically tailored to destroy acetaldehyde. This enzyme does to acetaldehyde what alcohol dehydrogenase does to ethanol: it removes a hydrogen atom, thus producing acetic acid, which is harmless. In a moment, we'll take a closer look at this second step in the body's detox process. But let's pause a moment and consider the key player in the first step—that garage-size molecule of alcohol dehydrogenase.

Why does this enzyme exist? Why do we have genes for building an enzyme specifically tailored to destroy alcohol?

Clearly, these genes did not just magically appear with the advent of the human discovery of fermentation. The answer to these questions lies in our guts. The helpful bacteria that populate our intestines often work under conditions similar to those experienced by yeast at the bottom of fermentation vats. Deprived of oxygen, these bacteria produce minute amounts of ethanol as a result of anaerobic metabolism. Apparently, natural selection favored creatures that could get rid of these tiny quantities of ethanol. Thus creatures making enzymes capable of destroying ethanol were more likely to survive than creatures that felt the full effects of internally produced alcohol. Not only does this explain why humans have the ability to sober up after drinking, but it explains why our enzymatic defenses are so easily overwhelmed: they were designed to handle only the minute amounts of ethanol secreted by bacteria in our intestines—not the comparatively massive quantities contained in even a single shot of scotch.

The biochemical protection of alcohol-destroying enzymes is not conferred equally among individuals, however. Alcohol dehydrogenase—like all enzymes—is manufactured according to blueprints stored in DNA. Since everyone's DNA is unique, individuals vary, sometimes strikingly, in the efficiency and activity of their alcohol dehydrogenase. One source of variation between people is a matter of sex. For reasons that remain unclear, alcohol dehydrogenase is less efficient in female stomachs than it is in male stomachs. In a recent study of this phenomenon, the enzyme activity of men was 70 to 80 percent greater than that of women (Frezza et al. 1990). This may account for the fact that, in general, women become intoxicated sooner in response to the same dose of alcohol than men do. It is also one of the reasons that the definition of "moderate drinking" is defined differently for men and women: two standard drinks a day for men as opposed to only a single standard drink a day for women (Gordis et al. 1995).

Prior to the study just mentioned, the different rates of intoxication between the sexes was thought to occur because women, on average, have a smaller volume of blood than men. An identical dose of alcohol, it was thought, would be more concentrated in women's blood than men's. But Frezza and his colleagues found that the difference between men and women in this particular area has much more to do with what happens in the stomach than in the blood. The subjects in this study were given ethanol either orally or intravenously. With oral administration, the ethanol was exposed to the alcohol dehydrogenase in the subjects' stomachs. Under these conditions, the researchers observed significantly higher blood levels of alcohol in the women compared with the men. But when the ethanol was given intravenously, *no* significant differences were found. The differences in total blood volume between the sexes, in other words, had no detectable effects on blood alcohol levels; exposure to stomach enzymes definitely did.

Interestingly, this heightened ethanol sensitivity in women appears to apply only to *young* women. Another study, done by German researchers, showed that the situation just described *reverses* in men and women over age fifty (Seitz et al. 1990). Alcohol dehydrogenase activity in men decreases significantly with age, to the point where the activity drops below that found in women of the same age. This means that men become increasingly sensitive to ethanol as they age, ultimately rendering them *more* sensitive than women.

As we'll see in a moment, sex isn't the only genetic factor involved in alcohol sensitivity. But since you're still in the stomach lining watching alcohol dehydrogenase go to work against ethanol, it's a good time to mention some important *nongenetic* reasons that people might vary in their response to a given dose of alcohol.

First of all, as many people know, food can affect the rate at which alcohol enters the bloodstream. This effect has less to do with the *type* of food eaten than with the *amount* eaten. When a sizable meal is consumed, the exit valve of the stomach—a muscular gate called the pyloric sphincter—closes so that the stomach can get to work digesting the food. This traps alcohol in the stomach, which not only prevents it from being rapidly absorbed in the small intestine but also increases the chances that it will be destroyed by the alcohol dehydrogenase found in the stomach lining. When the stomach is empty or contains only a small amount of food, however, the pyloric sphincter relaxes and allows alcohol to pass into the small intestine, which absorbs alcohol much more quickly than the stomach does.

A less commonly appreciated variable in an individual's response to alcohol involves aspirin. For reasons that remain unexplained, aspirin disables alcohol dehydrogenase (Risto et al. 1990). In one study, the average blood alcohol levels of subjects who consumed alcohol an hour after ingesting two Maximum Bayer® aspirin tablets were 26 percent higher than subjects who consumed ethanol *without* first taking aspirin. Clearly this represents a significant increase, particularly since many people assume that they can have one drink an hour and remain sober. Aspirin changes this equation—especially among women, for the reasons just mentioned. Other things being equal, taking aspirin will result in higher and longer-lasting blood alcohol levels. This impairment of alcohol dehydrogenase in the stomach can also be caused by certain drugs, such as the anti-ulcer drugs cimetidine and ranitidine.

Central Detox

Bored now with the spectacle of alcohol dehydrogenase ripping into ethanol molecules in the stomach lining, you extricate yourself and plop down into the flow of juices leaving the stomach. Soon you're through the pyloric sphincter and are sluicing into the upper reaches of the small intestine. As the wall of the small intestine comes into view, you notice that it looks furry. This "fur" explains why alcohol is so rapidly absorbed here, as opposed to its relatively slow absorption in the stomach. The walls of the small intestine are covered with millions of minute projections called villi. These structures give the small intestine a surface area of more than 200 square meters— roughly the size of a tennis court. This enormous surface area makes the small intestine an ideal place for the absorption of small molecules like water, glucose, amino acids—and alcohol.

Within seconds, you and the load of ethanol pass through the villi and are mixed with blood surging toward the liver. All blood from the digestive organs, including the stomach, is shunted to the liver, which filters toxins, bacteria, and other potentially harmful substances before they get into general circulation. Other than the brain itself, the liver is the organ most often associated with alcohol. And with good reason. Although alcohol affects every body system and every organ to some extent, the liver bears the brunt of the assault by virtue of its role as the body's detox center. Although the liver is amazingly resilient (remove half of it and it can regenerate fully), it is not invulnerable. Long-term exposure to alcohol can cause a number of crippling diseases, including cirrhosis—a permanent scarring caused by the death of liver cells.

As you enter the liver, you see ethanol molecules diffusing quickly out of the blood and into the surrounding liver cells.

You tag along behind one group of ethanol molecules. Dead ahead, you notice a familiar molecule: alcohol dehydrogenase. This one is slightly different from the ones you saw in the stomach. Amazingly, the liver deploys not one, but *seventeen* distinct varieties of alcohol dehydrogenase in its effort to defend the body against alcohol. Each variant has characteristics that differ subtly from those of the others, allowing the liver to detoxify ethanol at a broad range of concentrations and ambient conditions. Working full-tilt, these enzymes can intercept and disable all the roughly 200 quintillion ethanol molecules in a half-ounce of pure ethanol in about an hour. This fact is the basis for the one-drink-an-hour rule of thumb for remaining sober.

As we've seen, however, this rule must be applied carefully. It's most accurate for young, healthy males who slowly consume a modest drink over the course of an hour, who are taking no other drugs (such as aspirin) that interfere with the action of alcohol dehydrogenase, and who are not drinking on an empty stomach. Changing any of these variables means that more time must be allowed between drinks to ensure sobriety.

But even this stringent interpretation of the one-drink-an-hour rule is insufficiently narrow. It turns out that one's race, as it relates to the second step in the liver's detox process, can also affect one's sensitivity to ethanol.

Remember that acetaldehyde, the product of the first step of alcohol metabolism, is relatively toxic due to its propensity to bind with proteins. As just mentioned, in most people, this toxic acetaldehyde is rapidly converted into harmless acetic acid by an enzyme called aldehyde dehydrogenase. But because of a subtle change in the structure of aldehyde dehydrogenase, this second detox step is blocked in some people. Roughly half of all Asians produce an inactive form of aldehyde dehydrogenase (Chen and Yeh 1989). The mutation in the gene used to make the enzyme is astoundingly minor, given that aldehyde dehydrogenase is constructed from a string of 374 individual amino acids. *One* of those 374 amino acids is altered in the mutant, inactive form (Takeshita et al. 1993). That single change alters the physical shape of the enzyme enough to prevent the binding of acetaldehyde, thus rendering the enzyme impotent. When people with the mutant enzyme drink alcohol, their blood levels of acetaldehyde skyrocket, producing a symptom dubbed the "alcohol flush reaction." The drinker's face flushes bright red, and he or she experiences heart palpitations, dizziness, and nausea. All in all, it is reported to be a *very* uncomfortable experience.

In a curious twist of racial genetics, the problems caused by this inability to destroy acetaldehyde are exacerbated by the fact that many Asians *also* possess an unusually active form of alcohol dehydrogenase—the enzyme that *produces* acetaldehyde from ethanol. They are thus dealt a genetic double-whammy: they produce more toxic acetaldehyde than normal, but their ability to dispose of that acetaldehyde is severely limited.

Recent work has identified two subgroups of Asians with deficient enzymatic machinery. Those with nearly complete inactivation of aldehyde dehydrogenase (and thus those with the most pronounced reactions) have been termed "fast flushers," while those with only a moderately disabled enzymatic machinery have a less severe reaction and are called "slow flushers."

As unpleasant as it may be to experience, the cloud of the flushing response may have a silver lining. A study of 1,300 Japanese alcoholics found that *none* were fast flushers and only 8 percent were slow flushers (Higuci et al. 1994). The authors of the study conclude that the flushing response probably deters individuals from drinking—or from drinking enough to trigger alcoholism. The authors caution, however, that both general alcohol intake and alcoholism itself have been rising steadily in Japan in the past two decades. They suggest that environmental factors such as increased stress and easier access to alcohol owing to rising standards of living may be counterbalancing the genetic protection against alcoholism conferred by the flushing response, particularly among those with the less severe response.

As we will see in later chapters, genetically determined variations in alcohol metabolism are only one way that biology influences response to alcohol. But as the study just cited indicates, environmental influences can powerfully affect the way genetic predispositions are expressed in human behavior.

On the Loose

After passing through the liver, you exit into a nearby blood vessel. As you look around, you see that many ethanol molecules have escaped destruction. This isn't surprising, since the shot of scotch was administered in a single gulp on a relatively empty stomach—exactly the type of onslaught for which nature did not prepare us.

You are now headed, along with the load of ethanol, up to the heart and into widespread circulation through the body. And that means the end of our imaginary journey. Unlike the relatively constrained path from shot glass to liver, the paths taken by ethanol molecules once they leave the heart are so diverse that a single vantage point isn't very useful. But our journey has served admirably to introduce some specific actions of alcohol as well as some general operating principles of the body. We've "witnessed" the body's alcohol-destruction system in action—the enzymatic tag-team of alcohol dehydrogenase and aldehyde dehydrogenase—and we have seen that this system varies widely among individuals. We've also seen that gender, race, age, food intake, and the ingestion of drugs such as aspirin can affect a person's response to alcohol. You're now in a position to tailor the one-drink-an-hour rule of thumb for your own use.

Our microscopic vantage point has also allowed us to examine some of the body's basic molecular machinery. For instance, we met some ion channels—a class of molecules we'll meet again shortly because they are key players in the actions of alcohol and caffeine in the brain.

Finally, we've been reminded repeatedly that from nature's point of view, anyway, alcohol is a poison to be eliminated as quickly as possible. Which raises an obvious question: Why do humans have such a powerful urge to consume this poison? The answer lies in what happens when what's left of the shot of scotch finally reaches the brain.

> Why has wine the effect both of stupefying and of driving to a frenzy those who drink it?
> —*Aristotle, The Problemata*

Your Brain on Alcohol

3

Pharmacy in a Bottle

Alcohol has traditionally been called a depressant. The designation was made because at high doses, alcohol slows down the central nervous system. The classic symptoms of drunkenness—slurred speech, discoordination, diminished cognitive ability—arise from a depression of function in various parts of the brain. And alcohol is lethal at very high doses because it depresses nerve functioning in the brain stem to the point that breathing stops. But, as Aristotle observed more than two thousand years ago, alcohol does more than just "stupefy" those who drink it. It can also "drive to a frenzy." Intoxication can evoke boisterousness, talkativeness, aggression, ribaldry, and other behaviors more typical of a stimulant than a depressant. The standard explanation for these effects is that alcohol depresses the "higher" cognitive abilities, such as the ability to control emotions, thus allowing our more unruly, carnal sides to emerge.

Although it contains a grain of truth, this theory rather radically misses the mark. For one thing, it begs the question of how alcohol depresses brain function in the first place. But, more important, current research indicates that the central premise of the theory is wrong: alcohol is not simply a depressant that produces only *apparent* stimulation. In reality, alcohol *directly* stimulates the brain and exerts a host of more complicated effects as well.

It's true that, like ether, alcohol—especially at moderate to high doses—can act as a general anesthetic, depressing a broad range of central nervous system functions. But alcohol also mimics the action of the drugs cocaine, amphetamine, Valium, and opium (Charness et al. 1989; Koob and Bloom 1988; Weight et al. 1993). Like cocaine and amphetamine, alcohol directly stimulates certain brain cells. At low doses, it increases electrical activity in the same brain systems affected by these classic stimulants and can lead to feelings of pleasure and euphoria—feelings that may underlie much of alcohol's addictive potential. Alcohol also works on exactly the same brain circuits targeted by Valium; the calming, anxiety-easing effects of alcohol closely resemble those exerted by this famous tranquilizer. And alcohol also resembles opium because it can release our internal stores of the morphinelike compounds called endorphins, thus tapping into the brain's core pleasure circuits. In short, current research reveals alcohol as a far more complex and interesting drug than it was thought to be. It is a regular pharmacy in a bottle: a stimulating/depressing/mood-altering drug that leaves practically no circuit or system of the brain untouched. This broad scope, in fact, sets alcohol apart from many other drugs. Substances such as cocaine and LSD work like pharmacological scalpels, altering the functioning of only one or a handful of brain circuits. Alcohol is more like a pharmacological hand grenade. It affects practically everything around it.

Alcohol's lack of specificity makes it a somewhat maddening quarry for research scientists. It increases the firing of some nerve cells, or neurons, while decreasing the firing of others. It stimulates some regions of the brain while depressing others. And the effects it exerts can change with time and dose. Given this array of confusing variables, it is remarkable that anything is known about alcohol at all. But, in fact, an enormous amount has been learned, particularly in the past ten years. The current picture of what happens in your brain when you drink alcohol is often strikingly at odds with previous notions about how alcohol works.

Pruning

The simplest idea of how alcohol affects the brain is so entrenched in popular lore that it's the crux of a common joke about drinking: I'm not killing brain cells, I'm just pruning to allow for new growth. The idea seems to be that alcohol mows down neurons like so much Listerine®, killing them "on contact." Given what we've learned thus far about alcohol, this doesn't seem terribly far-fetched. After all, alcohol is a hazardous solvent excreted by yeast, and nature finds it sufficiently poisonous to equip us with an elaborate (if relatively ineffective) enzymatic defense system against it. In addition, very high concentrations of alcohol can indeed kill cells, which is why it is used as a disinfectant. But these lethal levels are never reached in the brain itself. Legal intoxication is reached when the concentration of alcohol in the blood reaches a mere .1 percent. That's a far cry indeed from the nearly 100 percent solutions used for sterilizing. Thus although alcohol does many things to the brain, one thing it clearly *doesn't* do is wipe out neurons indiscriminately.

Recently, Grethe Jensen and colleagues proved this fact by carefully counting neurons in matched samples of alcoholics and nonalcoholics (Jensen and Pakkenberg 1993). The samples were taken from people who had died of causes unrelated to drinking. When the two groups were compared, no significant differences in either the overall number or the density of neurons were found between the two groups.

As a statement about alcohol's action in the brain, then, the joke about pruning is wrong. But in one rather roundabout way, the joke contains an element of truth. It implies that the brain is so crammed with neurons that you can afford to kill off thousands with every drink without serious risk of dementia. This isn't far from the truth. The brain is a phenomenally complex piece of biological machinery. This is due in part to the sheer number of parts it contains: approximately 100 billion neurons and tens of billions of additional support cells called glia. One way to appreciate these numbers is to sneak up to them. It has been estimated that over the course of your life you'll lose roughly 7 percent of your brain's neurons from normal wear and tear (Dowling 1992). That 7 percent, translated into actual neurons lost, corresponds to an average daily loss of 200,000 brain cells. In other words, you are so fabulously endowed with neurons that you can afford to discard roughly a quarter-million of them each day without losing your mind—literally.

But as impressive as such numbers are, they aren't the real

reason the brain is so complex. After all, the liver is composed of billions upon billions of cells too, but it doesn't warrant the same superlatives. The brain's complexity derives from the way those 100 billion parts are wired up. Neurons communicate with one another by way of treelike extensions called dendrites and axons. Generally speaking, a neuron "listens" to incoming signals with its dendrites and "talks" to other neurons by sending signals out along its axon. A single neuron can communicate with as many as 50,000 other nerve cells in this way. And each of those connections—called synapses—is a vitally important link in the brain's information-processing abilities. In fact synapses, not whole neurons, are thought to be the fundamental unit of information storage and manipulation in the brain.

A better estimate, therefore, of the brain's true size can be found by multiplying 100 billion neurons by 1,000—a rather conservative average of the number of dendritic connections that each neuron makes with another. The result: 100 trillion synapses. 100 trillion functional units. As one neuroscientist quipped, "100 trillion synapses, hell, you can do anything with that. That's more than enough to contain a soul" (Johnson 1991).

But that's not all. The brain is not a printed circuit board. Connections between neurons change over time. New synapses form; old ones disappear. The brain's "wiring" is sculpted by experience. Information streaming in through our eyes, ears, and other sense organs can be captured because the connections between neurons can change in a split second, forming new circuits. Thus the true measure of the brain's complexity isn't just the raw number of neurons or dendritic connections in the brain at any given point in time, but the total *possible* number of connections—the number of ways neurons can be linked together into discrete patterns. This number cannot be reliably estimated, though it is widely viewed as far higher than 10^{78} (that's a 10 with seventy-eight zeros after it), which is a rough guess at the number of atoms in the entire universe (Hooper and Teresi 1987).

That's why you don't need to fret over your average daily loss of 200,000 neurons—it's virtually insignificant in the great scheme of things. But if alcohol doesn't actually kill neurons— if we've got so many that it probably wouldn't matter much if it *did* kill neurons—then what, exactly, *is* alcohol doing to all those cells? And why do chronic drinkers seem to show clear signs of cognitive dysfunction? As with the science of fermentation, it took a surprisingly long time for anybody to find answers to such questions.

Olive Oil Clues

In the late 1890s, two German scientists, E. Overton and H. Meyer, were investigating the ways that different alcohols dissolve in olive oil. As we've seen, ethanol is a small type of alcohol, with a backbone of only two carbon atoms. Methanol is smaller still, with but a single carbon. But larger alcohols have three, four, or more carbons. Overton and Meyer discovered that the length of an alcohol's carbon chain is related to its ability to dissolve in olive oil and other fats. Simply put, the longer the chain, the easier it is for an alcohol to mix with fat.

The interesting thing about this observation is that carbon-chain length is *also* related to an alcohol's anesthetizing and intoxicating powers. The longer the carbon chain, the more potent the alcohol as an intoxicant. This led to an obvious hypothesis: perhaps alcohol's power to intoxicate is related to its ability as a solvent—in particular, its ability to dissolve the fatty walls of cell membranes. Unfortunately for the Germans, the instrumentation of the day wasn't up to testing this hypothesis. They couldn't even *see* cell membranes, much less do careful experiments on them. And so for decades, their work with olive oil stood only as a tantalizing clue about how alcohol worked.

It wasn't until the 1960s and 1970s that the tools of neuroscience became subtle enough to pick up where Overton and Meyer left off. A wide range of experiments in those decades showed that, indeed, ethanol and other alcohols had a disordering effect on cell membranes—and the longer the alcohol molecule, the more disorder. Specifically, the alcohols appeared to fluidize membranes, to make them looser and easier to disturb.

As we saw during our microscopic tour of the tongue, cell membranes come studded with a wide range of protein molecules, such as the ion channels you watched opening and closing. Such proteins literally float in the membrane like so many icebergs, and so the fluidizing effect of alcohols could—at least theoretically—cause a wide range of difficulties. For instance, the movement of nutrients into a cell and waste products out of a cell might be disrupted, leading to impaired nerve-cell function. Until as recently as the 1980s, one or another version of this so-called lipid theory of alcohol's action held sway. Intoxication, it was strongly suggested, resulted from the disorganizing effect of alcohol on the membranes of nerve cells, leading to a depression in their functioning.

Then problems began to crop up. For one thing, the concentrations of alcohol used in the original experiments were far higher than those causing intoxication in humans. They were, in fact, lethal doses. Even more damning, the fluidization observed with realistic doses of alcohol was minor and could be duplicated simply by raising the temperature of the cells a few degrees above normal body temperature (Treistman et al. 1987). Since people with mild fevers do not become intoxicated, alcohol was clearly doing more in the brain than simply fluidizing neuronal membranes.

In light of these difficulties, attention has now shifted away from the membrane and toward the proteins embedded *in* the membrane. And the particular proteins now under scrutiny are the ion channels we met earlier. We've seen that by flexing or twisting (the exact motions are unclear), these large molecules can open or close a hole in the membrane. When open, the channels allow electrically charged atoms (ions) to surge into or out of a cell. This alters the cell's electrical properties.

Ion channels are found in the membranes of practically all cells, but they are particularly important for neurons. The traffic of ions into and out of neurons underlies their capacity to generate and transmit electrical signals.

A growing body of evidence shows that alcohol molecules directly affect the ability of ion channels to open or close. This basic action—not a general fluidizing effect on membranes— is now thought to be responsible for the range of behavioral phenomena lumped under the label "intoxication" (Weight 1992).

By revealing how alcohol interacts with ion channels, current research is shedding light on many ancient questions regarding alcohol's mode of action—not least of which is Aristotle's query about why alcohol seems to be both a stimulant and a depressant.

Alcohol as Ether

Alcohol's ability to depress brain function is one of its most obvious and potentially hazardous attributes. Recent evidence from a number of laboratories shows that one way alcohol achieves this depression is by interfering with a type of ion channel critical for the firing of neurons (Lovinger and Peoples

1993). Before proceeding, however, we need to consider briefly the widely used phrase "firing neuron." Since such firing lies at the root of all the many effects of alcohol—and caffeine, too, for that matter—understanding the basics of this business is rather important.

Neurons can "fire" because they generate a relatively large electrical charge across their membranes. In a sense, neurons are like microscopic batteries gradually storing charge, and then releasing it when fired. The electrical charge used by neurons is carried by ions—those electrically charged atoms mentioned previously. When positive and negative ions are mixed together, as they are in most parts of the body, their charges cancel out and the result is an electrically neutral solution. But if positive and negative ions are separated and concentrated, a charge difference is developed—a difference measured in volts.

Neurons achieve a separation of charge by forcibly moving ions on one side or the other of the fatty cell membrane, which is an excellent electrical insulator. Special protein molecules called ion pumps ferry positive ions out of cells, which leaves the outside of the cell positive and the inside negative.

The process of firing generally begins with incoming signals from other neurons. These "signals" are actually tiny surges of positive or negative ions entering the dendrites through ion channels. As a result, the electrical charge inside the neuron is in constant flux, moving above and below the average maintained by the ion pumps. In a very real sense, each neuron is performing a calculation: it is adding up the signals coming into it via the dendrites. If negative ions predominate, nothing happens. The cell is *already* negative and so adding *more* negative ions just pushes the cell farther in an electrically negative direction. But if large quantities of *positive* ions enter the neuron, the electrical charge produced by all those ion pumps is partially neutralized. If the cell's overall charge is neutralized below a certain critical point, a trigger is pulled causing a special class of ion channels to spring open near the base of the axon—the main fiber carrying messages away from the cell. These channels let in a flood of even *more* positive ions, which obliterates the electrical charge at that spot.

The sudden collapse of electrical potential around the base of the axon is "felt," in turn, by adjacent ion channels a bit farther down the axon. These channels now open. The charge collapses at this new point, triggering yet more channels to open farther down the axon. The process continues, like the flame of a firecracker fuse. This traveling wave of altered electrical potential is called an *action potential*, more commonly known as a nerve impulse. Action potentials are the stuff that minds are made of. Affect the way they are generated, transmitted, or received by other neurons and you affect the mind itself.

Action potentials zip down axons at about 225 miles per hour. When they reach the end of the axon, they don't automatically cause the next neuron in line to fire. Such an arrangement wouldn't be terribly effective, since a single firing neuron would quickly ignite a crippling chain-reaction of firing throughout the brain. To avoid this problem, all of the brain's billions of neurons are separated from each other by tiny, insulating gaps called synapses (Figure 3). To cross this gap, an action potential must be converted from an electrical signal to a chemical signal. It works like this:

When an action potential reaches the tip of an axon, it triggers the release of one or another kind of signaling molecule into the synapse. These signaling molecules are called neurotransmitters because they transmit messages between neurons. Neurotransmitter molecules are contained in tiny bags called vesicles inside the axon tip. When an action potential arrives, the bags rapidly fuse with the cell membrane, dumping their

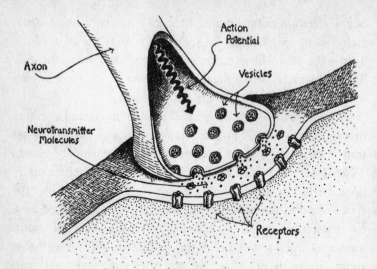

Figure 3. Neuronal synapse. (Ann Bliss Pilcher)

load of neurotransmitter into the synapse. In milliseconds, neurotransmitter molecules drift across the synapse and dock into specially designed receiving molecules on the "downstream" neuron. This docking is a matter of molecular geometry: the bumps and knobs of a neurotransmitter fit into corresponding dimples and holes on the surface of a receiving molecule called a receptor. Receptors come in dozens of varieties, each specially designed to accommodate one of the dozens of neurotransmitters used by the brain.

When the right neurotransmitter docks with the appropriate receptor, the physical structure of the receptor molecule changes. Sometimes receptors are ion channels, and the docking trips the channel open. Other times the receptor acts as a signaling station, triggering a chemical chain reaction inside the cell that sends a message to nearby ion channels to either open or close. Regardless of whether it's direct or indirect, however, the end result of neurotransmitters crossing a synapse is usually the flow of ions into (or out of) the receiving neuron. If the ion channel lets positive ions into the receiving neuron, the neuron is pushed toward firing. But if the receptor lets in *negative* ions, the downstream neuron is made more *resistant* to firing. Anything that interferes with these receptors influences the messages being sent from neuron to neuron in the brain.

Now it's time for a drink (figuratively speaking, of course).

Short Circuits and Blackouts

Having seen some of the mechanical underpinnings of brain function, we're able to appreciate how molecules of alcohol might plausibly affect this functioning. Let's start by looking at one way alcohol can slow down brain activity.

One of the major neurotransmitters used to send "fire" messages from one neuron to another is a molecule called glutamate. When glutamate is released into a synapse, it docks at a receptor that lets positive ions rush in. Since this makes it more likely that the receiving cell will fire, glutamate is called an *excitatory* neurotransmitter.

When you take a drink, alcohol molecules that escape destruction in the liver are quickly pumped up to the brain, where they infiltrate synapses everywhere. There they can bind to glutamate receptors. Nobody knows precisely *where* on the receptor alcohol binds, but it somehow warps the structure of the receptor just enough to interfere with its ability to open normally, thus muting glutamate's normal "fire" message. Alcohol's inhibition of glutamate receptors can be profound. After consumption of the equivalent of about two drinks in the space of an hour, glutamate receptor function can be reduced by more than 80 percent (Weight et al. 1993).

By inhibiting the brain's most common excitatory neuro-transmitter, alcohol effectively slows down activity in many parts of the brain. If the neurons in those areas control muscles, the inhibition can lead to relaxation and discoordination. If the neurons control speech, words slur and become increasingly imprecise. If the neurons control automatic bodily functions, heart rate and breathing are impaired. From a public health point of view, these are among alcohol's most dire effects. The inhibition of glutamate receptors is the molecular foundation of such grim statistics as the annual death of more than 20,000 people in alcohol-related traffic accidents.

The inhibition of glutamate receptors may also disable one of our most coveted intellectual capacities: the ability to learn.

Although it's often compared to a computer, the brain more closely resembles a tablet of wet clay into which impressions can be made, erased, and made again over time. This flexibility enables us to learn and remember. The current theory of memory suggests that you remember something when a specific constellation of neurons is stimulated vigorously. Whether it's a whiff of cinnamon or a catchy song, an incoming stimulus instantly lights up a particular constellation of neurons. If conditions are right, the connections between the neurons in the constellation are automatically strengthened in the process. If this pattern of neurons is stimulated in exactly the same way again, the network "lights up" more easily than it did previously. The original sensation is thus "stored" in these discrete patterns of tuned connections. The more often a particular pattern is stimulated, the more sensitive and permanent the connections between the neurons in the pattern become.

The technical term for such long-lasting changes in the strength of synaptic connections is long-term potentiation, or LTP. (A mirror phenomenon—long-term inhibition—is also likely to be involved in memory formation.) The discovery of

LTP in 1973 provided the first plausible mechanism to support the theory outlined earlier. When this phenomenon was investigated closely, it was discovered that LTP is blocked when a specific kind of glutamate receptor is disabled—the NMDA receptor.

Disrupting NMDA receptors has serious consequences. Rats, rabbits, and other animals injected with chemicals that block NMDA-receptor channels can't learn new tasks, such as negotiating their way through a maze, and they are incapable of forming new memories. Their abilities return when the effects of the chemicals wear off.

The salient point here is that, of all the glutamate-receptor channels (there are three basic types) the NMDA receptor is the *most* sensitive to alcohol (Weight et al. 1993). Experiments show a 30 percent reduction in LTP at alcohol concentrations reached after only a single drink (Blitzer et al. 1990). The impairment worsens with higher alcohol concentrations, stabilizing at roughly 80 percent with a concentration roughly equivalent to serious inebriation: a blood alcohol level of .2 percent—about twice the legal limit for intoxication in most states.

This research shows that alcohol—even at very low doses—disrupts the cellular machinery most widely believed to underlie our ability to form new memories. Since the disruption can occur at levels below those causing more obvious impairments of motor function and speech, people may not appreciate the degree to which their memories are being impaired.

Interestingly, the impairment is of the ability to form *new* memories, not the ability to recall stored memories. Intoxicated people who were asked to recall a list of words learned prior to intoxication showed no impairment of their recall ability (Birnbaum et al. 1978). In contrast, when the words were presented to people *already* intoxicated, their ability to recall the words later dropped significantly (Jones 1973). Results such as these suggest that the operations of memory acquisition and memory retrieval are separated in the brain and rely on different kinds of molecular machinery.

The memory impairment resulting from alcohol ranges from a barely detectable "cocktail-party amnesia" to the full-blown memory blackouts experienced by alcoholics. Inhibition of NMDA channels is the most likely cause of the moderate impairments, but the molecular basis for alcoholic blackouts has not been determined. It may be due to the combined effects of the inhibition of NMDA channels and the alteration of other types of ion channels that produce a massive inhibition of nerve-cell firing in the hippocampus, a portion of the brain critical to memory formation.

If alcohol affected only neurons and neural networks that rely on the neurotransmitter glutamate, it would still be a powerful substance. But such effects alone would not make for a very popular drug. Indeed, people drink alcohol *despite* the fact that it depresses overall brain function and can radically interfere with the ability to learn. Accounting for alcohol's enormous popularity and explaining its myriad other effects requires that we look beyond glutamate to some of the brain's other important neurotransmitters.

Liquid Valium

Anxiety is an unpleasant emotional state that differs from related states such as fear, aggression, and confusion. Not only does anxiety *feel* different, but at a purely neurological level, it *is* different. Evidence for this comes from experience with a family of drugs called benzodiazepines, of which Valium is the most well known. At low to moderate doses, these drugs significantly reduce anxiety without impairing or disrupting other brain systems.

Valium works by enhancing the function of a receptor that plays yin to glutamate's yang in the brain. Instead of passing on a message to "fire," this receptor makes it *harder* for a neuron to fire. The receptor in question is triggered by a neurotransmitter known as gamma-aminobutyric acid, or GABA. When GABA docks at its receptor, the associated channel opens and lets *negative* ions rush in, which pushes the cell even farther from its trigger point for firing. Although such inhibition might seem counterproductive, it is actually crucial. Normal brain function depends on both excitatory *and* inhibitory neurotransmitters. The situation is analogous to the operation of an automobile, which requires both an accelerator and a brake. Glutamate is one of the brain's accelerators, and GABA is one of its brakes. We've already seen that by interfering with glutamate channels, alcohol interferes with the accelerator, making it harder to gain speed. Now we'll see that another way alcohol slows the brain is by *increasing* the sensitivity of the brakes. This, in fact, is how Valium and other benzodiazepines work. These compounds bind to GABA receptors, which alters their shape and makes them three times more sensitive to GABA molecules (Ashton 1992). Valium, in other words, makes the brain's natural "brake" three times stronger.

Alcohol, like Valium, can reduce anxiety (at least in the short term), and it accomplishes this in exactly the same way that Valium does: by binding to GABA receptors and enhancing their function. As with glutamate, the exact binding site of alcohol hasn't yet been pinpointed. But one thing is clear: it's different from the sites used by Valium and other benzodiazepines. That's why it's so dangerous for people on anti-anxiety drugs such as Valium to drink alcohol. When both alcohol *and*

Valium molecules bind to the GABA receptor, they warp the channel to a much greater degree than does either drug acting alone, producing a correspondingly larger inhibition of neuronal firing. Ignorance or disregard of this synergistic behavior can be fatal.

The fact that alcohol mimics the calming actions of Valium is widely seen as part of its attraction as a drug. Anxiety is so omnipresent in today's society that it is hardly surprising that people turn to a readily available, legal, and relatively inexpensive anxiety-reducing drug for relief.

And yet alcohol intoxication involves more than mere relief from anxiety. As Aristotle noted, alcohol also appears to exert directly positive effects: stimulating, euphoric, pleasurable feelings not accounted for by its ability to banish anxiety or induce a general sedation. The experience of alcoholics also indicates that alcohol does more than relieve pain—though that may certainly be a component of its addictive quality. Alcohol induces a powerful craving for a kind of pleasurable feeling not attributable to its effects on glutamate and GABA ion channels. Understanding these latter effects requires a brief look at the brain circuits of bliss.

Raiding the Pleasure Center

It is critically important for animals to discriminate between behaviors that enhance their chances of survival and behaviors that undercut those chances. Animals possessing some kind of internal compass to help them make these choices would clearly have an advantage over animals forced to learn which behaviors are adaptive and which aren't. It is not surprising, therefore, that brain circuitry has evolved to perform this function.

In the broadest sense, there are two such systems: rewarding circuits and punishing circuits. Like a biochemical carrot and stick, these systems generate pleasurable or painful feelings that powerfully guide behavior. The reward circuits generate cravings that impel an animal toward such things as eating, drinking, and procreating. When one of these actions is completed successfully, neurons in a specific part of the brain release chemicals that elicit feelings ranging from a calm satiety to orgasm.

The existence of a discrete reward center in the brain was first demonstrated in 1954 by physiologist James Olds, who placed very fine electrodes in the brains of rats and allowed the animals to stimulate certain areas of their brains by pressing a lever in their cage. He found a dramatic response when the electrodes were placed in a region called the mesolimbic area. Rats with electrodes in this area seemed to enjoy the stimulation very much and worked very hard to obtain it, even if this meant learning to negotiate a complex maze. When allowed to stimulate themselves at will, they would sometimes do so at the rate of over a hundred times a minute for hours on end. In fact, some animals starved themselves rather than give up pressing the lever.

The electrodes gave the rats access to their own stores of bliss-producing neurotransmitters. Normally meted out frugally and only after the accomplishment of some important survival-enhancing task, these compounds were now available at the press of a lever. Allowed free access to their own reward centers, many of the rats became hopeless lever addicts. Of course, implanted electrodes aren't the only way to stimulate the brain's mesolimbic reward center. Cocaine, heroin, amphetamine, nicotine, and a great many other drugs give humans a lever for accessing their pleasure centers.

The role of the mesolimbic system in addiction and drug use is so compelling that it is the focus of intense research. The system is extremely complex, and we are far from a complete understanding of its function. Yet, thanks in part to research on drugs such as cocaine and opium, some of the basic neurotransmitter systems vital to the reward center have been elucidated.

Alcohol very likely affects *all* the neurotransmitters used in this center. To date, two of these have received particular attention: dopamine, and the opiumlike endorphins. By altering these neurotransmitters, alcohol is now thought to evoke the euphoric, hedonic sensations associated with intoxication. These are also the brain chemicals widely regarded as most intimately involved in alcohol's high potential for abuse among certain drinkers.

Alcohol modestly increases dopamine levels in the reward circuits of the brain, making it a weak cousin of cocaine and amphetamine (Di Chiara and Imperato 1988). This release of dopamine is thought to underlie the initially stimulating, energizing feelings often experienced by drinkers.

The "high" one gets from alcohol is, of course, quite different from that achieved by cocaine and amphetamine. These drugs are much more potent and are practically surgical in their effects. They zero in on dopamine while leaving other neurotransmitters untouched. The stimulation produced by alcohol, in contrast, is modest to begin with and must compete with the simultaneous depressant effects caused by the inhibition of glutamate channels and the enhancement of GABA channels.

Alcohol's effect on dopamine levels has been found to be most pronounced in the first twenty minutes of exposure (Friedman et al. 1980, Frye and Breese 1981). This may explain why the early stages of intoxication feel qualitatively different from later stages. In animals, the initial "hit" of dopamine from ingestion of alcohol correlates with a brief increase in activity, which then declines to levels below that displayed before the alcohol was administered.

Exactly how alcohol boosts dopamine levels isn't known. Alcohol might act directly on dopamine receptors, making them more sensitive than normal in a manner analogous to the sensitization of GABA receptors. Alternatively, alcohol might act indirectly by affecting neurons impinging on the reward center, rather than acting on the reward center itself. Whatever the mechanism, alcohol's ability to activate the dopamine circuits in the brain's reward centers provides the first good explanation for its stimulating effects since Aristotle commented on these effects two thousand years ago.

The other mediators of pleasure being actively investigated are endorphins, the body's natural painkillers. During times of severe stress or injury, endorphin molecules are released from the pituitary gland and block pain messages arriving from various parts of the body. Secondary to this important task, endorphins also trigger the release of dopamine in the brain's mesolimbic reward center, which, as we've seen, directly elicits pleasurable feelings. Endorphin release is thus doubly rewarding: it dampens pain and produces, via dopamine, a mild "high."

Given what we've already learned, it probably won't come as a surprise that alcohol has been found to trigger the release of endorphins from the pituitary gland (Gianoulakis et al. 1990). If this were alcohol's *only* effect, drinking it would produce a subtle "high" similar to that felt by marathon runners and other athletes who come by their endorphins naturally. As it is, drinkers experience an endorphin boost simply as one of many elements in a very potent mix—yet another dimension in the subjective experience of intoxication.

Alcohol thus resembles opium and its derivatives morphine and heroin, all of which target the endorphin system. Alcohol is much less potent than opiates, however, because it works in an entirely different way.

Opiate molecules fit snugly into the molecular receptors designed for endorphins. They are, essentially, fake endorphins. Opiate users thus can give themselves an "endorphin rush" far more intense than anything possible with only their own natural supply of these pleasure compounds. But ethanol molecules don't look anything like endorphin molecules. They aren't fake endorphins at all. All alcohol can do is tap into one's existing store of endorphins. Since no new endorphins or endorphin look-alikes are added to the system, the opiate-like high achievable with alcohol is limited. As with dopamine, the precise mechanisms behind alcohol-induced endorphin release aren't yet known.

An Unfinished Picture

This chapter has presented a portrait of alcohol's actions in the brain that is startlingly different from that commonly held. Far from being a simple depressant, alcohol is a subtle, complicated drug that exerts a wide range of pharmacological effects.

We've seen that by inhibiting glutamate receptors, alcohol induces a general sedation and significantly impairs the brain's ability to store new memories. By increasing the sensitivity of GABA receptors, alcohol mimics Valium and reduces anxiety. Like a weak version of cocaine or amphetamine, alcohol boosts dopamine levels, producing a brief period of heady stimulation. And by releasing endorphins, alcohol resembles opium, giving users a rush of pleasure similar to the "natural high" experienced after a vigorous workout.

As comprehensive as this list is, however, it probably does not tell the full story of alcohol's effects on the brain. The brain uses at least forty neurotransmitters, which act on more than one hundred types of receptors. Scientists are still learning how alcohol alters the way many of these transmitters and receptors work, and some of these alterations may turn out to be just as significant as the ones considered here.

One important research subject is the neurotransmitter serotonin, the target of the widely used antidepressant drug Prozac. By boosting serotonin levels, Prozac can alleviate depression, enhance motivation, and increase self-confidence. Preliminary studies suggest that alcohol also acts on the serotonin system. It has been found, for instance, that moderately high doses of alcohol increase the electrical current associated with one type of serotonin receptor by almost 60 percent (Lovinger and Peoples 1993). This increased current is functionally equivalent to boosting levels of serotonin in the synapse—exactly what Prozac does (Weight 1994).

Another line of research linking alcohol and serotonin involves rats bred for their avid preference for alcohol. The brains of these rats have significantly *lower* serotonin levels than do the brains of rats that don't crave alcohol. One hypothesis: the former group may like alcohol so much because it helps compensate for their genetically faulty serotonin machinery. In humans, some alcoholics have lower amounts of serotonin breakdown products in their cerebrospinal fluid than do nonalcoholics, indicating that their brains are manufacturing less serotonin than normal.

Finally, Prozac and similar serotonin-boosting drugs have been shown in several studies to modestly reduce alcohol consumption among both alcoholics and nonalcoholics (Amit et al. 1984; Lawrin et al. 1986). These observations have led to efforts to use Prozac as a supplement to more traditional alcoholism treatment programs.

Despite such findings, alcohol is clearly different from Prozac. For one thing, whereas Prozac often *alleviates* depression, alcohol almost always exacerbates the feelings of hopelessness and inertia associated with clinical depression. For another, it generally takes four to six weeks for Prozac's positive effects to "kick in," whereas alcohol's effects are felt very rapidly. At the moment, therefore, too little is known about the molecular and neurological mechanisms of serotonin to say anything definitive about how alcohol is related to that system.

Serotonin is just one of the neurotransmitters under investigation as neuroscientists continue to explore how alcohol works in the brain. New sites of action are sure to crop up as this investigation continues. But the molecular mechanisms we've explored in this chapter go a long way toward explaining some of the age-old mysteries of drinking and intoxication.

> rink, Sir, is a great provoker of three things . . .
> se painting, sleep, and urine. Lechery, Sir, it
> vokes and unprovokes; it provokes the desire,
> t takes away the performance.
> —The Porter, in William Shakespeare's Macbeth

Sex, Snores, and Stomach Aches

Beyond the Brain

We now know that alcohol is a much more interesting drug than the simple depressant it is commonly thought to be. It produces complicated, often paradoxical effects in the brain that mix and overlap with one another to create an equally complex intoxication. Drinking alcohol can push one's mental state in practically any direction, from a stimulated, energetic euphoria to a dark, brooding hopelessness.

But the brain is hardly the only organ affected by alcohol. It's just the one most *obviously* affected. The alcohol in a shot of liquor, a glass of wine, or a bottle of beer infiltrates every nook and cranny of the body, and it provokes changes in other organs that can be just as complex and paradoxical as those in the brain.

Most people, of course, drink alcohol for its mind-altering properties. But that's not the only reason people reach for a bottle. For instance, many a homemade cold remedy contains alcohol in one form or another, owing to the belief that a small dose has restorative powers. Alcohol is also widely taken as a "nightcap" to induce sleep. Alcohol—particularly red wine—is downed these days with the thought that modest consumption confers protection from heart disease. And alcohol is undoubtedly the world's most widely used aphrodisiac.

But does alcohol really enhance the pleasures of sex? Is it good for the heart? Will it help you sleep better or recover from a cold more quickly? As with alcohol's effects in the brain, the answers to these questions are both more interesting and less clear-cut than is commonly thought.

Le Paradoxe Français

In January 1996, the United States government unveiled new guidelines for a healthy diet. Prepared by a committee from the Department of Agriculture and the Department of Health and Human Services, the guidelines drew attention because, for the first time, they acknowledged the positive health benefits of moderate alcohol consumption. "Alcoholic beverages have been used to enhance the enjoyment of meals by many societies throughout human history," the report noted, "and accumulating evidence suggests that moderate drinking may lower the risk of heart attacks." This stance contrasts sharply with the line taken in the 1990 guidelines that "drinking has no net health benefit." The guidelines do not encourage people to drink—and the recognition of alcohol's health benefits are carefully hedged with warnings that *no* alcohol should be consumed by children, teenagers, women who are trying to conceive or who are pregnant, and anyone planning to drive or who cannot control drinking to moderate levels. Still, the announcement was a dramatic and highly public endorsement of a sizable body of medical research into the health benefits of alcohol in general, and red wine in particular.

Such studies began in earnest in the 1970s when scientists began taking note of the so-called French paradox. In studies at that time comparing the diets and rates of illness in different countries, the French—who consume a lot of cheese, cream, meat, and other high-fat foods—had one of the *lowest* incidences of major heart disease. Only the Japanese—whose diet is high in rice, fish, and other low-fat foods—had less heart disease. Since France has a high per capita consumption of wine, researchers began to look at whether the two facts were related. The results from more than a decade of study strongly suggest that, in fact, there *is* a connection. It now appears that daily consumption of one or two standard alcoholic drinks reduces the risk of heart disease. The evidence also suggests that red wine, in particular, is good for the heart.

Does this mean that doctors are now prescribing wine for patients at risk for heart attacks? Hardly. The issue has provoked carefully worded articles in medical journals about the proper advice to give patients who ask about wine's benefits. For instance, an editorial in the *New England Journal of Medicine* notes that "there now seems little doubt that alcohol exerts a protective effect against coronary heart disease" (Friedman and Klatsky 1993). Yet after reviewing the complexities of the issue, the editorial stops short of a recommendation that patients begin drinking. "As in other areas of health care," the authors demur, "the patient must, with our guidance, make the final decision."

Several well-designed studies of heart disease have found that people who drink no alcohol have a slightly higher lifetime risk of coronary heart disease than people who consume light to moderate amounts of alcohol. But when consumption rises, risk rises as well—to levels much higher than those faced by abstainers. Early studies showing this risk curve were faulted for including among the teetotalers people who recently quit drinking. But more rigorous studies eliminating such people from the survey and controlling for other potentially confounding variables such as diet and smoking came to the same conclusion: moderate drinking appears to reduce the risk of atherosclerosis (clogged arteries) and myocardial infarcts (heart attacks). Some studies calculated a risk reduction of as much as 50 percent; a more conservative figure, derived from a number of similar studies, is roughly a 35 percent reduction (Friedman and Klatsky 1993).

Of course, it is one thing to find an association between two variables; it is quite another to prove that alcohol *causes* a reduction in heart disease. Making this leap requires a satisfying explanation of how alcohol can have a palliative effect on the cardiovascular system. Three possible explanations have been put forward, each backed up by solid research and none likely to be the sole mechanism. As our discussion of the brain revealed, alcohol affects nearly everything it touches. So it would not be surprising if it conferred its benefits on the heart by altering several things at once.

One of the earliest theories has also been one of the most contentious. Initial studies showed that alcohol boosted levels of high-density lipoproteins (HDLs) in the blood. HDLs are considered "good" because they transport cholesterol from the blood to the liver, where it is either transformed or destroyed. The more HDL, in other words, the greater your body's ability to move cholesterol out of your arteries. In contrast, *low*-density lipoproteins (LDLs) work in somewhat the opposite manner, transporting cholesterol *away* from the liver and out to peripheral tissues. Although this function is just as critical to health as that performed by HDLs, LDLs have been labeled "bad" since at high levels they often dump their loads of cholesterol on artery walls, where it can accumulate into sticky, clogging plaques.

If alcohol raises HDL levels, as the initial research suggested, it could plausibly explain why alcohol appears to be so "heart healthy." But then, as so often happens in science, more research complicated the picture. It turns out that there are several kinds of HDL. At first, it was thought that only one kind, HDL_2, was beneficial, and alcohol apparently did not affect HDL_2; instead, it seemed to work on HDL_3. It has taken years to sort the whole thing out, but researchers now believe that not only are *both* HDL_2 and HDL_3 important in reducing the risk of heart disease, but alcohol raises the levels of both. Thus the research has come full circle. Alcohol's effect on HDLs (both kinds) is once again thought to account at least in part for its beneficial effects on the heart.

Meanwhile, other researchers were focusing on red wine, rather than on all alcoholic beverages. One of the reasons for this attention was that people in countries such as Scotland, Finland, and the United States who consume more of their alcohol in the form of beer and hard liquor have higher mortality rates from heart disease than do the French and Italians, who consume most of their alcohol in the form of wine, particularly red wine. The research conducted to date has turned up several intriguing results.

John Folts and his colleagues at the University of Wisconsin

Medical School have found convincing evidence that a class of compounds collectively called phenols dramatically reduces the ability of blood platelets to clump together into clots. Specifically, Folts found that two kinds of phenols—quercetin and rutin—abolish or significantly reduce the clot-forming ability of platelets in dogs. Red wines contain much higher percentages of phenols than do white wines; when Folts tried the experiment with white wine, he found little effect. Thus one way that red wine, at least, may help is by reducing the formation of potentially deadly arterial clots and clogs.

The French paradox might also be related to recent findings from Queen Elizabeth Hospital in Birmingham, England. Researchers there were intrigued by findings that red wine, in test-tube experiments, was found to be an antioxidant. Oxidation is simply the reaction of oxygen with some other compound. Fire is an example of rapid oxidation, while rust is indicative of slow oxidation. In your body, oxidation of glucose is essential for energy production. But sometimes oxidation is not so helpful. When some compounds are oxidized, they become reactive and unstable. In this state, they can easily damage or disrupt nearby molecules. One molecule prone to this kind of damaging oxidation is low-density lipoprotein. Oxidized LDL reacts with proteins and other compounds in cells, interfering with normal functioning.

Which brings us back to red wine. The Queen Elizabeth Hospital researchers knew that red wine blocked the oxidation of LDL in the test tube, but they wanted to know if it did the same thing in human beings. In an experiment with five men and five women, they found that the levels of antioxidants in the blood rose considerably after volunteers ate a meal and drank two 5-ounce glasses of Bordeaux red wine. When the same meal was consumed without red wine, antioxidant levels actually fell.

Which—if any—of these three mechanisms is the one primarily responsible for the French paradox remains to be seen. It may be that all three contribute to the overall effect.

With all this evidence suggesting that moderate consumption of wine—and probably other forms of alcohol as well—confers protection against heart disease, why isn't everyone reaching for their favorite bottle of cabernet? There are several reasons. First of all, the estimated 35 percent reduction in risk is offset to some extent by slight *increases* in the risk of contracting one of the many diseases associated with drinking alcohol, or of being involved in an alcohol-related accident. The French, while enjoying their much reduced rates of heart disease, develop liver disease at a rate that is roughly twice that of Americans (Dolnick 1990). In addition to taxing the liver, moderate drinking has been associated with a slightly increased risk of breast cancer and cancer of the bowel. And, of course, even a single shot of liquor consumed quickly can produce transient blood alcohol levels high enough to reduce reaction times and impair coordination, thus increasing the risk of accidents.

Second, advising abstainers to begin drinking could lead to increased alcoholism because it is not yet possible to predict who will succumb to alcohol's addictive potential. And finally, drinking modest amounts of wine or other types of alcohol is hardly the only way to reduce the risk of heart disease. Other methods, such as losing weight, quitting smoking, and exercising, offer even greater benefits and have fewer associated risks.

The moral of the French paradox is that if you don't drink, don't start just to help your heart—it's not worth it. If you drink moderately, current research suggests that you needn't worry that you're hurting your heart—in fact, you're probably helping it. And if you drink *more* than two or three drinks a day, you should probably cut back, if for no other reason than drinking this much isn't going to help your cardiovascular system and will probably compromise your overall health.

And I in My Cap . . .

A lot of people have a hard time getting to sleep at night—roughly 30 million people in the United States alone, according to one estimate (Palca 1989). Stress, emotional upheaval, depression, medications, not to mention excess caffeine can all leave you staring at the ceiling until the wee hours of the morning.

No one knows how many insomniacs turn to the nearest bottle of booze in an effort to get some sleep, but the mere fact that a late-night nip is known as a "nightcap" suggests that this is a common use of alcohol. At first glance, this seems to make sense. Many people have felt sleepy after drinking and, as we've seen, alcohol depresses some brain circuits by enhancing the activity of GABA receptors. In fact, the most widely prescribed class of sleeping pills are benzodiazepines such as Halcion, Valium, Xanax, and Restoril, which work by enhancing GABA-receptor functioning.

But alcohol actually makes a rather poor sleeping pill. It may, indeed, nudge you into dreamland, but you don't necessarily stay there and you have a very good chance of waking in the morning feeling decidedly un-rested.

As we've seen, alcohol is both a depressant *and* a stimulant. Among other things, it boosts dopamine and endorphin levels, both of which can elicit stimulating or mildly euphoric sensations. These effects are particularly pronounced at relatively low doses—just the kind of doses typical of a "nightcap." Thus a single shot of whiskey or a small glass of wine taken just before bedtime may have an effect that is exactly the reverse of the one being sought by the drinker.

More significantly, controlled sleep studies have shown that volunteers who consume moderate to heavy amounts of alcohol before going to bed tend to drop off to sleep relatively quickly, but then wake often in the middle of the night and have difficulty going back to sleep (Stradling 1993). The best explanation for this is the so-called rebound effect, otherwise known as acute tolerance.

Tolerance is the adjustment of the brain to the presence of a drug, necessitating larger and larger doses to achieve the same effects as the original dose. Tolerance has been observed in humans with every drug of abuse, including alcohol and caffeine. The tolerance that develops from long-term, repeated exposure to a drug is called chronic tolerance, and it's the basis for many of the severe problems associated with drug addictions. But tolerance can also develop after as little as a single dose of a drug (Iversen and Iversen 1981). In the case of alcohol, for instance, the brain can adapt very rapidly, changing within hours to counteract the resulting imbalances. Such tolerance is relatively short-lived, but it's enough to disturb sleep in some important ways.

One way is by disrupting an important phase of sleep nicknamed REM, which stands for rapid eye movement. The brain's electrical activity during REM sleep looks almost identical to that observed when a subject is wide awake. Heart and breathing rates are highly variable, the eyes move rapidly under closed lids as if the person were watching a movie, and, if awakened, subjects often report that they were dreaming. People typically have between four and six periods of REM sleep a night and spend about 25 percent of their total sleeping time in this stage.

Alcohol consumption tends to reduce the amount of time a person spends in REM sleep. The reasons for and importance of this finding remain unclear, but REM sleep has been found to be critical for learning specific kinds of new tasks (Karni et al. 1994)—possibly because during REM sleep the brain "replays" the day's events in a way that solidifies and consolidates learning and memory. This effect of REM sleep seems particularly important for so-called procedural memory, which is what we use when we learn to ride a bike or touch-type. In the Karni study, subjects deprived of REM sleep—but not other phases of sleep—made no progress in learning certain procedural tasks, whereas volunteers allowed REM sleep but deprived of non-REM sleep *improved* their performances overnight.

It is likely that non-REM sleep plays an important role in memory and learning as well, though in different ways than REM sleep. Both types of sleep, however, are disrupted by alcohol. During the early phases of the night, alcohol reduces REM sleep. Later on, however, the rebound effect leads to restlessness, which interferes with both REM *and* non-REM sleep.

The disruptive effects of alcohol on sleep can be exacerbated by caffeine. Caffeine is broken down by the liver much more slowly than alcohol. It takes about five hours for the liver to metabolize half of a given dose of caffeine (Stradling 1993). This has some interesting implications for people who attempt to use alcohol as an antidote to the wakefulness induced by too much caffeine. The sedating effects of a moderate to strong dose of alcohol may, at first, override the stimulating effects of the caffeine, promoting sleep. But by the middle of the night, the alcohol will be metabolized and the rebound effect will set in. Meanwhile, the caffeine will *still be in circulation*, which exacerbates the mild stimulation resulting from the rebound effect. The result is a biochemical double-whammy that can leave you awake in the middle of the night and groggy the next morning, which sets the stage for even *more* caffeine drinking and a perpetuation of the cycle the following night.

To Your Health?

The use of alcohol as a restorative medicine is as ancient as its use as an intoxicant. For instance, the New Testament book of Timothy contains the suggestion to "drink no longer water but use a little wine for thy stomach's sake and thine infirmities" (1 Timothy 5:23).

When the art of distillation was discovered in the Middle Ages, the potent extract that resulted was deemed both a good medicine in itself and an ideal base for the creation of other remedies by the addition of herbs and other ingredients. In fact, the original name for alcohol was *aqua vitae*, Latin for "water of life." It was regarded as a life-giving, life-affirming liquid.

The idea that alcohol had medicinal qualities persisted well into the twentieth century. Even during America's periodic fits of prohibition, the members of temperance societies were asked only to forswear alcohol as a beverage, not as a medicine (Tice 1992). Physicians in the eighteenth and nineteenth centuries administered many of their medicines by dissolving them in wine made from grapes, elderberries, blackberries, or apples. Rye whiskey, mixed with rock-sugar syrup, remained a popular cough remedy into the early twentieth century. In the South, many believed that mint julep prevented malaria. And in the mid-nineteenth century, a wide range of patent medicines with alcohol contents averaging around 40 percent (about the same as scotch) were bought by millions seeking cures for everything from baldness to gout. One particularly popular brand, Lydia Pinkham's Vegetable Compound, was 20 percent alcohol and carried on each bottle the slogan "Trust Lydia Pinkham, not the doctor who doesn't understand your problems." The tonic was marketed principally to housewives and grossed approximately $300,000 in one year, making Pinkham one of the richest women of her day.

Surprisingly, Pinkham may have been on to something. Even though the alcohol content of patent medicines such as Pinkham's was relatively high, the amount generally taken was small (at least for those who really took the substances as medicine), and recent research suggests that low doses of alcohol may confer some immunological benefits. Before we take a look at these intriguing findings, however, it's worth pointing out the stark differences between acute (short-term) use and chronic use. Practically all the news from research into the immunological impacts of chronic alcohol use is negative. Heavy drinkers, in addition to greatly increasing their risks for everything from liver disease to high blood pressure, are also far more susceptible to infectious diseases. Additionally, long-term use of alcohol depresses the immune system by inducing malnutrition, vitamin loss, and general incapacitation of the liver. These effects can leave a heavy drinker vulnerable to a host of diseases ranging from cirrhosis to cancer, and serious brain disorders such as the memory-destroying Korsakoff's syndrome.

There is little doubt, therefore, that heavy drinking is bad for one's health in general, and bad for one's immune system in particular. There is quite a lot of doubt, however, about the effects of light to moderate drinking. The evidence here is mixed. For instance, studies of alcohol's effects on natural killer cells—the "scavenger" cells that destroy virus-infected and cancerous body cells—have produced uneven results. One study showed that natural killer cell activity was suppressed in mice that ingested high doses of alcohol over a two-week period

(Meadows et al. 1989). But another study, using low doses of alcohol, showed that alcohol actually *enhanced* natural killer cell activity (Saxena et al. 1981).

The most intriguing recent study comes from the Common Cold Unit of Britain's Medical Research Council (Cohen et al. 1993). This study is significant because it was unusually well controlled and involved a large number of research subjects: 417. Volunteers were given in-depth physical exams, and were questioned extensively about their smoking and drinking habits as well as other pertinent aspects of their lifestyles. They were also given personality tests, since some studies have found an association between susceptibility to infections and personality type. Blood samples were drawn for analysis, and then most of the subjects were given nose drops containing an infectious dose of the virus that causes the common cold. (As a check, twenty-six subjects were given saline drops. As expected, none of these people got colds.)

To ensure maximum control, volunteers were quarantined at the Cold Unit for seven days after being given the nose drops. During this time, they were monitored daily for cold symptoms, blood samples were drawn, and a wide range of other measures were taken to assess their reactions. Subjects who normally smoked or drank were allowed to continue doing so throughout their quarantine. Because of the large number of study subjects and the elaborate measures taken to ensure accuracy and control, it took nearly three years to conduct the study.

As the authors of the study note in their report, the results pertaining to alcohol were "unexpected." Contrary to what previous studies on chronic use of alcohol had led them to expect, these researchers found that alcohol significantly *increased* resistance to infection. The volunteers who drank alcohol contracted fewer colds during the study period than did the volunteers who drank no alcohol. And, in general, the more alcohol consumed, the fewer colds volunteers contracted.

A limitation of this study is the small number of volunteers who drank heavily. In fact, only 10 percent of the entire sample and 5.6 percent of the nonsmokers downed more than an average of three drinks daily. The authors concluded that they couldn't say anything about what happens to cold risk as drinking becomes heavy. As just noted, however, evidence from research on alcoholics indicates that heavy drinkers are *more* vulnerable to infections than are light and moderate drinkers or abstainers.

As for the mechanism by which alcohol could confer protection from colds and help reduce cold symptoms, there are only educated guesses at the moment. One possibility is that alcohol somehow limits the reproduction of viruses, either directly or via an enhancement of the body's immune response. Another possibility is that alcohol inhibits the production of histamines—the compounds responsible for runny noses and some other unpleasant symptoms of colds. These and other possibilities are now the focus of ongoing research.

The protective effects of alcohol seen in this study were not strong enough, however, to overcome the effects of smoking. As expected, smokers had the highest incidence of colds, and the more volunteers smoked, the more likely they were to catch a cold—regardless of the amount of alcohol they consumed.

As intriguing as these results are, much remains unclear about the impact of low doses of alcohol on the immune system. The authors of the study end their article with a warning that their results should not be taken as a suggestion that nondrinkers begin to drink. As with heart disease, the results simply indicate that those who already drink moderately are apparently not *increasing* their risk of getting a cold. Such people may indeed be giving themselves some small margin of immunological protection.

Sex, Lies, and Alcohol

In 1976 forty undergraduate men at Rutgers University in New Jersey set aside their modesty for science. The men took part in a study designed to shed light on the paradox observed by Shakespeare that alcohol "provokes the desire, but it takes away the performance." Previous research had shown that by most physical measures alcohol is bad for sex. Scientists had measured how alcohol affected penile swelling, vaginal engorgement, time required to achieve orgasm (during both intercourse and masturbation), and vaginal lubrication. The results were strikingly uniform: alcohol inhibited all these responses. Erections were slower to rise and quicker to fall, vaginas were slower to lubricate, and orgasms were slower to arrive. The mechanics behind these reductions in sexual response are still not clearly understood. It is strongly suspected that alcohol inhibits firing in the peripheral nervous system. That includes the nerves terminating at the penis, the clitoris, and the vagina.

The penis and clitoris would respond to this inhibition in very similar ways because, anatomically speaking, they are nearly identical except for size. Both structures contain spongy tissue that can swell and become erect during sexual arousal. Both penis and clitoris are usually flaccid because the arteries supplying these organs with blood are under most circumstances clamped tight. Sexual arousal creates nerve impulses that *relax* these arteries, allowing blood into the spongy tissue and causing it to swell. Sexual arousal is thus fundamentally dependent on relaxation, not tension. Conversely, orgasm involves contractions of a variety of muscles, which just goes to show that sex is as complex at an anatomical level as it is at a behavioral level.

It might seem natural to assume that since alcohol is a muscle relaxant, it would facilitate sexual arousal by relaxing those all-important penile and clitoral arteries. But note that the arteries just mentioned relax in response to the *firing* of nerves, not the inhibition of firing. Those firing nerves can originate in the brain, as when a person has an erotic fantasy, or from the lower spinal cord, as when the genitals are directly stimulated. The key is that anything that blocks this firing will block the arterial relaxation needed to achieve an erect penis or clitoris. This is exactly how alcohol is thought to dampen sexual response, though many details in the process remain obscure.

There are times, of course, when a little dampening is precisely what a drinker is aiming for. Masters and Johnson (1986) estimate that between 15 and 20 percent of American men have at least some difficulty controlling premature ejaculation. They casually note what is largely ignored by most books dealing with sexual problems: that many men find they can retard overly rapid ejaculation with a judicious dose of alcohol prior to sex. Of course, it's easy to overdo it. If a man drinks *too* much alcohol, he may well find his sexual response retarded to the point of impotence.

The effects of alcohol just described, however, tell only a small part of the story. That's because in humans, sex can be a good deal more complicated than a matter of conjoined genitals. The brain gets into the act as well. And therein lies a tale.

Despite scientific evidence that at a purely physical level alcohol retards sexual response, many people report that moderate amounts of alcohol are *good* for sex. In one of the largest surveys addressing the issue, 45 percent of men and 68 percent of women said that alcohol *enhances* their sexual enjoyment

(Athanasiou et al. 1970). The answer to this apparent paradox lies in the old joke about the brain being the body's most important sex organ. Especially where alcohol is concerned, this is no joke at all—as the men from Rutgers ably demonstrated in the 1970s. That study involved an effort to separate the *physical* effects of alcohol from the effects of a person's *belief* about the consumption of alcohol—something that required a bit of judicious deception on the part of the scientists (Wilson and Lawson 1976). It worked like this.

The male students were randomly assigned to one of four groups. One group was given vodka and tonic, and was *told* that it was vodka and tonic. Another group got just tonic, and was also told the truth about what they were drinking. The third and fourth groups, however, were lied to. The third group got vodka and tonic, but was *told* that they were drinking plain tonic water. (The vodka in these volunteers' drinks was mixed in a 1:5 ratio which was undetectable.) The last group was *told* that they were getting alcohol, but in fact they were given tonic water in glasses smeared with a few drops of vodka to produce an alcohol smell.

The ruses were remarkably effective. Questioned after the experiments, not one of the volunteers who were duped said they suspected anything unusual.

Following the ingestion of their drink (either alcoholic or nonalcoholic), the volunteers were outfitted with a variety of monitors to gauge temperature, heart rate, and penile swelling. Each volunteer then watched an erotic video. After the film, the volunteers were interviewed extensively and were informed of the truth if they had been in one of the two groups that were misinformed about their alcohol intake.

The results were striking. The subjects who *thought* they drank alcohol were most highly aroused—whether they actually drank alcohol or not. The men who thought they drank alcohol and who actually *got* alcohol were the most highly aroused. The men who thought they got alcohol but got only tonic water were slightly less aroused, but these men were significantly *more* aroused than those who expected tonic water but actually drank alcohol. Thus it was the *belief* in alcohol consumption that proved significant to sexual response, not the presence or absence of alcohol. The belief overcame any of the physiological dampening effects that the alcohol might have had.

These findings have not been the only ones showing that belief and expectations are more important than purely physical effects of alcohol. A similar study found that people who thought they drank alcohol were significantly more aggressive in a social situation than people who thought they drank tonic water—regardless of whether they actually did or did not drink alcohol (Lang et al. 1975). More recent studies have shown that belief similarly effects female sexual response and the response of sexually inhibited males (Lang et al. 1980). When it comes to alcohol, in other words, people often feel what they expect to feel. This process can obviously be self-reinforcing. The experience of an enhanced sexual encounter under the influence of alcohol can lead to increased expectations of similar results the next time around.

One recently reported interaction between alcohol and sex deserves mention, if only because it has been so badly explained in the popular press.

In a 1995 study, Finnish and Japanese researchers who were studying some metabolic aspects of alcohol consumption unexpectedly discovered a relationship between alcohol consumption and testosterone levels in women. They found that the ordinarily low testosterone levels in women rise dramatically one to two hours after women imbibe alcohol, and that the rise was most dramatic in women who were either ovulating or taking oral contraceptives. The finding was published in the staid scientific journal *Nature*, but it was picked up by many considerably less-reserved media outlets because testosterone has been shown to increase sexual desire in both men and women. Many media reports suggested that women might thus respond more favorably to sexual advances after a few drinks.

This is quite a leap from the very carefully drawn findings of the study. For one thing, the findings have yet to be replicated. More important, it remains to be seen whether the levels of testosterone detected in the study have anything to do with actual behavior. There are good reasons, in fact, to suspect that they do not. First of all, the study simply measured total testosterone in the blood of female volunteers. It turns out that most of a woman's testosterone is not biologically active—it is bound to a protein in the blood and does nothing. Whether alcohol ingestion actually raises the level of biologically active testosterone is unknown. And, as we've just seen, the beliefs and expectations about alcohol's effects are likely to be more powerful influences on behavior than any effect exerted by testosterone.

It is important to remember that these results—as well as all the results showing the impact of expectation on sexual response—are based on *moderate* drinking. The blood alcohol levels among the volunteers in the expectancy studies, for instance, were equivalent to what would be found after the consumption of only two or three standard drinks over the course of an hour. (Again, a standard drink is defined as a half-ounce of pure alcohol—the amount generally found in a 12-ounce can of beer, a 5-ounce glass of wine, or a 1.5-ounce shot of whiskey.) If higher doses had been used in the experiments, it is likely that all the physical variables measured, including penile swelling, would be adversely affected. There are limits, in other words, to mind over matter. Given a high enough dose, all the expectation in the world won't rouse a penis or clitoris anesthetized by alcohol.

Indeed, the harmful effects of chronic drinking have been so exhaustively chronicled that it's easy to see why those in positions of influence are cautious in their comments on such a matter as the potential usefulness of low doses of alcohol for premature ejaculators. A brief list of such consequences will suffice to illustrate the point. Chronic alcohol use can quash the libido of both men and women. It can seriously depress testosterone levels in males, it often causes a pronounced shrinking of the testicles, and it strongly impairs the ability to achieve and sustain an erection. In women, chronic use of alcohol reduces vaginal response, and it can cause irregular menstruation and induce premature menopause.

The Exception to the Rule

Related to the issues of alcohol and sex is the matter of alcohol's dire effects on a developing fetus. Although some doctors continue to believe that a very occasional drink by a pregnant woman is harmless, the tide of medical opinion on this matter has shifted in recent years in light of new research. For instance, although it has long been known that alcohol passes quickly and easily across the placental barrier between mother and fetus, it now appears that alcohol may affect an embryo even before it has implanted itself in the uterine wall and become engaged with the mother's blood (Coles 1994). Much research is now aimed at discovering the exact mechanisms by which alcohol may harm a developing fetus. More than likely, alcohol impairs many critical molecular systems at once.

One system now being scrutinized is absolutely essential for the proper wiring up of neurons in the brain of the developing fetus. Special molecules called adhesion molecules guide the migration of developing neurons and help them to make stable connections to other neurons. Michael Charness and his colleagues at the Harvard Medical School demonstrated that alcohol "strikingly reduces" the ability of certain adhesion molecules to promote the formation of stable multicellular organizations (Charness et al. 1994). The disruptive effects of alcohol on cell adhesion molecules is a specific example of the havoc alcohol is thought to wreak on many other molecular mechanisms critical for the proper development of a fetus.

Although it takes very heavy drinking indeed (an average consumption of forty-two standard drinks a week) to produce the severe physical deformities and mental retardation characteristic of fetal alcohol syndrome, *no* safe threshold has been found for the far more subtle and difficult-to-measure effects of alcohol on mental development. To quote from a recent study of the matter: "For some behaviors, such as mental development, even the smallest prenatal dose of alcohol appears to have some adverse effect on the fetus, and the severity of the effect increases gradually with increasing levels of exposure" (Jacobson and Jacobson 1994).

This inability to determine a safe level of drinking has led many doctors and public health officials such as the U.S. surgeon general to advise women who are either pregnant or trying to get pregnant to abstain completely from alcohol. And it's probably not a bad idea for the male to quit drinking too—at least while trying to father a child. New evidence shows that alcohol can have an adverse effect on sperm, which may induce subtle yet marked deficits in the offspring of alcohol-exposed fathers (Cicero 1994).

In sum, if you are having sex with the intention of conceiving a child, most current research suggests that you stick to sparkling cider and other nonalcoholic drinks. If you're simply having sex for the pleasure of it, the research indicates that moderation is prudent. Generally speaking, the impact of alcohol on sexual response is dose-dependent. At light to moderate levels, the brain is more important than the alcohol: the response you feel will have more to do with what you *think* you will feel than with the pharmacological impact of alcohol on your sex organs. That means alcohol can either help or hurt sex, depending on your expectations. As alcohol consumption increases so does its power to override the mind and directly dampen sexual response. And, at high doses, as Shakespeare's Porter observed, you may well find your desire "provoked," but you will most likely be robbed of your performance.

The Morning After

Hangovers are the bane of drinkers. The throbbing head, nausea, irritability, dry mouth, lethargy, and hypersensitivity to light and sound make this condition so unpleasant that most people try to avoid it at all costs. Aside from the obvious tack of drinking slowly and in moderation, there are innumerable folk remedies aimed at avoiding or curing hangovers. Most of these ideas are ill-founded, some are downright harmful, and a few actually provide some relief.

The pounding headache common to hangovers has two possible sources. First of all, as Shakespeare's Porter again pointed out, alcohol is a diuretic—that is, it promotes urination. It does this by blocking an important substance in the kidneys called antidiuretic hormone, or ADH, which adjusts the porousness of the microscopic tubes that carry urine out of the kidney. Normally, most of the water in urine is recycled through the porous walls of the collecting tubes. But when ADH release is blocked by alcohol, the tubes become *less* porous, thus cutting down on reabsorption and increasing urine output. (As we'll see later, caffeine also promotes urination, but in an entirely different way.) Somewhat ironically, then, drinking large amounts of alcohol can lead to mild dehydration. This, in turn, can lead to both a dry mouth and a headache owing to reduced blood pressure in the cranial vessels. The second way alcohol can induce a headache is by relaxing and enlarging the same vessels in the head—an action that compounds the low-blood-pressure problem created by dehydration.

Both of these problems are, to a certain extent, correctable. Downing plenty of water both during and after drinking can help prevent dehydration (though it may add to a restless night by increasing your need to go to the bathroom). Drinking a caffeine-containing beverage in the morning may help also because caffeine constricts cerebral blood vessels. Taking an aspirin or two may help also, though only by masking the pain, not by solving the basic problem. Taking aspirin *before* you start drinking is *not* a good idea. As noted in Chapter 2, aspirin interferes with alcohol dehydrogenase, which can lead to higher blood alcohol levels and *worse* hangover symptoms.

Another common hangover symptom is a general lethargy and muscular weakness. One factor contributing to this malaise is the buildup of lactic acid in the muscles that can follow heavy drinking. As most athletes know, the fatigue and cramping caused by strenuous exercise result from the accumulation of lactic acid and a subsequent disruption of the acid–base balance in the muscles. Drinking alcohol can do roughly the same thing, though by a different route. The enzymatic destruction of alcohol in the liver requires many important "helper" molecules. But these helper molecules are normally used to process many other toxins, including lactic acid. When presented with a load of alcohol, the liver and all its enzymatic machinery drop what they're doing and go to work on the alcohol. This allows less dangerous toxins like lactic acid to accumulate, creating overly acidic conditions in your muscles.

The body's acid–base balance can be thrown off another way as well. The principal product of alcohol metabolism is acetic acid, which is useful in many ways. But produced in excess from the breakdown of alcohol, it can simply acidify the blood, exacerbating the lethargic feelings produced by the lactic acid buildup. There is little one can do—other than wait—to correct such imbalances. Mild exercise may help a little by increasing blood circulation and thus flushing lactic acid from the muscles. But this strategy could easily backfire, since strenuous exercise would simply produce *more* lactic acid, making the situation worse.

The queasy stomach common to hangovers is often attributable to the increased acids secreted by the stomach in response to alcohol. Of course, other factors may be at work as well, not least of which could be episodes of vomiting, which would leave the stomach both empty and overly acidic. Eating a light carbohydrate such as toast, crackers, or cereal is often recommended in this situation because it helps neutralize acid and is easily digested.

Hypersensitivity to light or sound may be due to the "rebound" effect discussed earlier in this chapter. A heavy bout of drinking will produce temporary withdrawal symptoms as the brain and body strive to rebalance themselves. Since withdrawal generally produces symptoms that are the antithesis of the original effects of a particular drug, the rebound from alcohol often brings with it increased excitability, depressed mood, and sensitivity to stimuli.

The fact that a hangover is, in part, a drug-withdrawal symptom accounts for the long-standing "hair-of-the-dog" cure. (The phrase comes from an old British saying: "A hair of the dog that bit you"—a metaphor for the idea that a small amount of the same kind of liquor drunk to excess the previous night will cure a hangover.) This "cure" can actually work—temporarily. By re-creating the conditions to which the brain and body had become accustomed during the night of drinking, a "hair-of-the-dog" nip of alcohol can bring some relief. But, of course, this only postpones the final reckoning and leads to more intense withdrawal symptoms later on. Fortunately, many people are dissuaded from a "hair-of-the-dog" because they have a natural aversion to alcohol following a significant pub crawl.

Many people feel that different kinds of drinks produce different kinds of hangovers. From a purely theoretical point of view, there is some logic to this. Wines, liquors, and beers contain hundreds of complex molecules that give these drinks their characteristic flavors, smells, and appearances. Such compounds are collectively called "congeners." Generally speaking, the more congeners a drink has, the darker it will appear and the richer it will taste. Red wine, for instance, has more congeners than white wine. Scotch, cognac, and brandy have more congeners than gin, and gin has more congeners than vodka, which is arguably the most congener-free liquor of all.

The problem with congeners is that there are so many of them that nobody has gone to the trouble of testing to see what, if any, effect they have on either intoxication or hangovers. (A few congeners, as mentioned, *have* been tested, but for their effects on the cardiovascular system, not on hangovers.) One congener with proven abilities to contribute to intoxication is methanol, ethanol's simpler cousin. Methanol isn't a direct product of fermentation—it is probably derived from the breakdown of pectins in fruit-based wines or liquors. Red wine has more methanol than white wines, but even the amount in red wine is so small (less than 1 percent by volume) that it is unlikely to play a role in hangovers. Other congeners include dozens of phenols, tannins, sulfur-containing compounds, organic acids, amino acids, esters, sugars, and gasses such as carbon dioxide. In all, more than five hundred distinct kinds of congener molecules have been identified in wine alone (Amerine and Roessler 1983).

It is possible that some people are more sensitive to the congeners in certain drinks than in others, and thus find that those drinks (such as red wine) give them worse hangovers. It is equally plausible that congeners have nothing to do with hangovers, and any appearance to the contrary is due either to expectations on the part of the drinker or simply to the very real effects produced by the alcohol.

The Middle Road

In this chapter we've explored some of alcohol's effects on the human body. We've seen that moderate amounts of alcohol, particularly red wine, confer protection against heart disease—though doctors don't recommend that those currently abstaining begin drinking to secure these modest benefits. We've seen that alcohol can at first promote sleep, and then disturb it by inducing a stimulating "rebound" effect. We've looked into the age-old idea that a measured dose of alcohol helps speed healing or prevent sickness and found a grain of truth in an otherwise unproved notion. We've been reminded that the most important organ in sexual response is the brain, and we've seen that alcohol's inhibiting effects on the sex organs become increasingly important as the dose increases. And finally we've explored the miseries of hangovers and seen how little can be done to prevent one, other than drinking moderately in the first place.

Indeed, moderation has been a repeated refrain in this chapter. The research to date on the wide-ranging effects of alcohol on the human body—both positive and negative—indicates that for moderate drinking, the net health impact is minimal or even slightly beneficial. Moderate drinking, remember, is usually defined as no more than two standard drinks a day for men and no more than one standard drink for women (Gordis et al. 1995). Remember, too, that this rule of thumb doesn't apply to pregnant women or to men and women who are trying to conceive a child.

And the biggest caveat, by far, to the rule of moderation is that all of the findings presented here apply *only* to people not at risk for alcoholism.

The best current estimate is that roughly one in ten drinkers is alcoholic. That means that for one in ten drinkers, the concept of moderate drinking is a dangerous illusion. Although some therapy programs claim that some alcoholics can resume moderate, controlled drinking, there remains great debate over this suggestion and substantial doubt about its practical implications.

It is to the subset of drinkers for whom alcohol is powerfully addicting that we now turn. No other area of alcohol research is as charged with emotion, contention, and debate. And yet the recent discoveries here are among the most fascinating in the entire field.

Some are fast and some are slow
Some are up and some are down
Not one of them is like another
Don't ask us why,
Go ask your mother.

—Dr. Seuss, *One Fish, Two Fish,*
Red Fish, Blue Fish

Demon Rum

5

Of Mice and Men

Since you can't open up a person's skull and probe around after
they've had a drink to see what's happening, scientists who are
curious about such things use animals. Rats and mice are by
far the favorite creatures for alcohol research. They're relatively
inexpensive, and their brains are a lot like human brains, only
smaller.

But as valuable as animals are, they have a serious drawback:
by and large, they hate alcohol. When alcohol studies using
animals began in earnest in the 1950s, it was found that when

dogs, cats, primates, rats, and mice were given a choice between an alcohol solution and water they almost invariably chose water. This posed quite a problem for researchers who wanted to see how alcohol intake—particularly long-term intake—affected the brain and other body systems.

Researchers overcame their subjects' natural aversion to alcohol in a variety of ways. They administered alcohol intravenously, for instance, or they disguised the alcohol with sugar or sugar substitutes. Some scientists even filled the animals' cages with vaporized alcohol, thus using the respiratory system as a means of alcohol ingestion. As awkward as such methods sound, they nonetheless allowed researchers to learn a great deal about how alcohol works. Still, humans obviously don't inhale their drinks (at least not literally), and they usually drink voluntarily. Indeed, the situation that many researchers were most interested in—deliberate, chronic drinking—was the one most difficult to accurately model with naturally abstinent lab animals.

Then, in the 1950s, scientists in Chile, Finland, and the United States independently succeeded in breeding strains of mice that actually *liked* alcohol (Crabbe et al. 1994). In 1959 the team of Gerald McClearn and David Rodgers at the University of California, Berkeley, discovered that a strain ingloriously called C57BL/Crgl clearly preferred alcohol to water when given a choice (McClearn and Rodgers 1959). They drank solutions with an alcohol content of 10 percent—roughly the same as white wine, and strong enough to make them tipsy.

The discovery was welcomed by experimentalists. They now had an animal that, like some humans, preferred alcohol to water. But the existence of alcohol-drinking mice was more profound than a mere methodological breakthrough—it struck to the heart of a central issue in alcohol studies. These mice were impossible-to-deny evidence that a preference for alcohol could be due to *genes*. The C57BL/Crgl mice were not some exotic species of mouse with radically different physiologies from other mice. They were simply one of six carefully bred families of *Mus musculus*—common house mice—tested by McClearn and Rodgers. The situation is analogous to selecting a few members of six unrelated human families, giving them two unlabeled glasses of liquid, letting them sample both, and finding that the members of one family preferred their water laced with alcohol.

The mice's preference was compelling because it clearly had nothing to do with upbringing, culture, peer pressure, stress, expectations, advertising, emotional trauma, or any other variable that can influence human drinking. The preference shown by the C57BL/Crgl mice was *internally* generated. They inherently liked the taste, the intoxication, or some other quality of the alcohol. Subsequent breeding bore this out. When alcohol-preferring mice were bred together, the inborn predilection strengthened: the grandchildren of the original mice drank more and drank higher concentrations of alcohol than their grandparents. Likewise, when mice that avoided alcohol were bred, their progeny became *less* willing to take even a sip of the hard stuff.

When these experiments were conducted, the science of genetics was in its infancy. Nobody knew what was going on inside those mice, though it was clear that the answer could be found somewhere in the tight coils of their DNA. Today, of course, genetics is one of the hottest fields in science. Genes now can be "read" with relative ease, and this new ability is revolutionizing our understanding of both physical and mental illnesses. Specific genetic defects have been found responsible for Huntington's disease, sickle-cell anemia, cystic fibrosis, muscular dystrophy, and a host of other diseases. And since DNA directs the construction of brains as well as bodies, genetic variation is coming to be seen as a key player in people's mental makeup.

This new appreciation for the way genes can subtly influence things like personality and mood has contributed to a sea change in the popular view of alcoholism. Alcoholism was long thought to be caused by a failure of will, a lack of moral fiber, or simple irresponsibility. The pendulum of public opinion has in recent years swung in the opposite direction: most people now view alcoholism as a disease caused by a genetic malfunction that renders sufferers predisposed to abuse alcohol. A 1987 Gallup poll found that nearly 90 percent of Americans believe that alcoholism is a disease, and more than 60 percent think that it may be inherited.

Evidence to support this model of alcoholism has been accumulating for decades. The discovery of those alcohol-preferring mice in the 1950s was one of the early contributions to what is popularly known as the "disease model" of alcoholism. The heady enthusiasm generated by the early findings of genetic predispositions to alcoholism, however, has faded and been replaced by a growing appreciation for the limitations of a purely biological approach to problem drinking. For one thing, despite forty years of searching, nobody yet knows why those mice discovered in the 1950s like alcohol. Whatever the genetic difference is between the tippling mice and their non-drinking brethren, it is so subtle or complicated that it has yet to be identified. But even if the genetics responsible for the mice's preference *were* identified, it is not clear that those findings would really shed much light on human drinking. As similar as mouse brains are to human brains, mice are not men. It turns out that the human brain's capacity to generate things like beliefs and values can rival the power of genes to influence drinking behavior.

Within the field of alcoholism research there is a growing

appreciation for the subtlety of the disorder—fueled primarily by the fact that virtually all the evidence amassed to support a role of genes in alcoholism indicates that faulty genes cannot be the *only* cause of alcoholism. Culture, family environment, learning, stress, even the hoary old notion of willpower, can play important roles as well. In other words, it is no longer nature versus nurture. It's nature *and* nurture. Or, perhaps more accurately, nature/nurture—not two separate entities, but two sides of the same coin.

In this chapter we will survey this new and still-evolving conception of alcoholism. We'll see what is known about why mice and men vary in their propensity to drink. In the process, we'll see that the once seemingly clear line between "alcoholic" and "nonalcoholic" has become quite blurry. No longer is it simply a matter of having or not having a disease. The forces at work on an alcoholic are at work, to one degree or another, on everyone, including the abstinent. The science of alcoholism is inexorably leading toward a view of the problem that is more complicated, more human, and more honest than either of the polar extremes that have characterized the debate for centuries.

Papa's Legacy

Alcoholism tends to run in families. The prevalence of alcoholism in the general population of males is estimated to be between 3 and 5 percent, while the prevalence among male relatives of alcoholics is about 25 percent (Goodwin 1985). For females, the figures are much lower, though the trend is similar. The prevalence of female alcoholism in the general population is between .1 and 1 percent, while the prevalence among female relatives of alcoholics is estimated at between 5 and 10 percent.

These numbers say nothing about the causes of alcoholism. Many things in addition to genes get passed from generation to generation, among them learned behaviors such as might account for a tendency to drink. The numbers just cited also strikingly contradict the notion that children of alcoholics are somehow *destined* to become alcoholics themselves. Although having a close alcoholic relative (parent or sibling) clearly increases one's risk of alcoholism, it is equally clear that having such a relative does not, by any means, condemn one to alcoholism. Seventy-five percent of males with an alcoholic relative do *not* become alcoholics themselves; more recent figures put the figure closer to 80 percent. Between 90 and 95 percent of women with an alcoholic relative escape the disorder. If alcoholism is a disease, in other words, it either is inherited in a most peculiar manner or is so weak that most people manage to overcome it.

One way researchers have tried to tease apart the entwined strands of genetic and environmental influences on alcoholics is to study twins. There are, of course, two kinds of twins. Identical twins form from a single egg and thus share identical genes, while fraternal twins are derived from two separate eggs and are no more genetically similar than non-twin siblings. If alcoholism truly has a genetic component, then identical twins should tend to develop more similar drinking patterns and problems than fraternal twins.

Similarity in a trait is measured with a value called concordance. If twins are identical in a trait—eye color for instance—the concordance is 100 percent. If they are completely divergent in a trait, the concordance is zero. Identical twins, clearly, are far more concordant in general than are fraternal twins.

The results of numerous studies from around the world clearly show that both genes and the environment influence drinking behavior. As many suspected, all of the studies have found that identical twins share the trait of alcoholism or problem drinking more often than either fraternal twins or completely unrelated people (Department of Health and Human Services 1993). For instance, a recent study of female twins found that the heritability of alcoholism was between 50 and 60 percent (Kendler et al. 1992). The authors concluded that "genetic factors play a major role in the etiology of alcoholism in women." But the *magnitude* of the differences observed in the heritability of alcoholism in twin studies is often surprisingly modest. A study of identical male twins showed a concordance rate for alcohol abuse or dependence of 76 percent, while that of fraternal twins was 61 percent. Although this is a statistically significant number, it's hardly a ringing affirmation of the disease model. The figures for women were even less impressive: a 36 percent concordance for identical twins compared with a 25 percent concordance for fraternal twins (Pickens et al. 1991).

This is an example of how a single study can support diametrically opposed views: those who favor genetic influences can point to the differences in concordance as proof, while those who think environmental conditions play a bigger role note that if alcoholism were purely a matter of genes, then the concordance for identical twins should be close to 100 percent. That it's *not* indicates that environmental variables are at work.

In reality, few people argue this way. Although it is a point often lost in lay discussions of alcoholism, practically nobody in the field believes that the disorder results from either genes or the environment alone. Even the most die-hard champions of genetics acknowledge that no amount of genetic predisposition can induce alcoholism if, for example, no alcohol is available for consumption. Likewise, even those who think upbringing or some other environmental factor is key to alcoholism admit that for at least some people a genetically based vulnerability may play a role. The one thing that seems clear— at least to scientists—is that alcoholism is *not* the result of a single faulty gene such as that underlying sickle-cell anemia or Huntington's disease. Most researchers tracking down the genetics of alcoholism and other types of addiction now assume that these traits spring from the combined influence of several genes, not one.

Some of the most interesting evidence to support this idea comes from studies on those alcohol-preferring mice discovered in the 1950s.

Mouse Tales

If you wanted to, you could pick up the phone today and order a mouse or a rat that displays any of more than a dozen distinct reactions to alcohol. You could, for instance, buy a mouse that becomes sleepy and motionless after a modest dose of alcohol. Such mice are called long-sleep (LS) mice because it takes them an unusually long time to wake up from an alcohol-induced nap. Conversely, another strain of mice gets drowsy on the same dose of alcohol, but quickly returns to normal. Hence, they're called short-sleep (SS) mice.

Or, if you'd prefer, you could choose between a COLD mouse and a HOT mouse. The metabolism of COLD mice slows down following a drink of alcohol, while the metabolism of HOT mice speeds up. There are also mice that become energized and active after drinking alcohol, and their opposites that become lethargic. There are even two strains that exhibit opposite sensitivities to withdrawal symptoms: a strain that displays the tremors and seizures typical of humans in the throes of delirium tremens (DTs) and a strain resistant to such symptoms.

The moral of this story is that, to a certain extent at least, the different physiological aspects of intoxication can be selectively bred into or out of animals. The anesthetizing aspect can be separated from the stimulating aspect, for instance, or sensitivity to withdrawal can be separated from the thermal effects of alcohol ingestion. It appears, therefore, that independent genetic factors influence alcohol sensitivity, tolerance, dependence, and self-administration and that these traits are supported by distinct neurobiological mechanisms (Crabbe et al. 1994). This may be one reason for the diversity in the experiences of drinkers. With an unknown number of genes at work helping shape one's sensitivity to alcohol, it's not surprising to see variations in people's responses—and that doesn't take into account the even greater variations produced by people's different personalities and other traits.

Some attempts to categorize the different responses to alcohol have been made. The most widely accepted distinction is between type I and type II alcoholism. Type I is more common, appears in both men and women, is less severe than type II, and often appears in midlife rather than early on (Cloninger et al. 1981). Several studies suggest that this type of alcoholism has a less pronounced genetic component and that its expression is strongly dependent on environmental influences. Type II alcoholism, in contrast, is characterized by a severe susceptibility that seems to be expressed regardless of the environment. Type II alcoholism occurs only in men, develops early (often in adolescence), and is much more difficult to treat than type I alcoholism.

Cloninger's division of alcoholism into two types is not the last word on the matter. Others have suggested a third type: alcoholism arising from a primary antisocial personality or mood disorder that exists prior to drinking (Schuckit 1985). The extent to which these categories reflect underlying genetic differences is unknown. If, as the animal studies suggest, aspects of intoxication such as craving and withdrawal sensitivity are regulated by different genes, then people with defects in such genes could plausibly express different forms of the disorder.

This idea is simply a hypothesis at the moment, though it is being taken very seriously by a number of researchers. The animal studies are so tantalizing that many teams of scientists are sifting through animal genomes looking for the kind of definitive defect or alteration they suspect lies at the bottom of the behavioral variations that can be bred in or out. Recently, they've had some good luck. The resistance to alcohol shown by short-sleep mice has been linked to a tiny variation in the gene used to build a type of GABA receptor in mouse brains. This variation, which slightly changes the structure of the receptor, is apparently vital to making the receptor sensitive to alcohol. Without this piece, alcohol can't alter the GABA receptors as it usually does, thus leaving the mice relatively resistant to intoxication.

As exciting as this kind of discovery is, it's a long way from a definitive statement about how genes affect human drinking behavior. Thus far researchers have been unable to find a human equivalent of the mouse GABA variant, much less a gene responsible for alcoholism in general. Thus despite widespread popular belief in the genetic origins of alcoholism, the scientific jury is still far from reaching a verdict in the case. Nothing illustrates this fact so compellingly as the story of what was, at one time, the leading candidate for the putative "alcoholism gene."

D_2 or Not D_2?

It began with a highly publicized press conference arranged to coincide with the appearance of the April 14, 1990, issue of the *Journal of the American Medical Association* containing an article by Kenneth Blum, a pharmacologist at the University of Texas, and Ernest Noble, a psychiatrist at the University of California. The pair announced that they had identified a genetic defect in the dopamine system of some alcoholics. Dopamine, as we saw earlier, is a key component of the brain's reward pathways—in many ways, it is *the* neurotransmitter of pleasure.

Blum and Noble examined DNA samples from the brains of corpses of thirty-five alcoholics and thirty-five nonalcoholics. They found that a variant of the gene for a specific type of dopamine receptor (D_2) was present in 69 percent of the alcoholics but only 20 percent of the nonalcoholics. They claimed that this variant, called the A1 allele, results in fewer D_2 receptors in the brain (Noble et al. 1991). Fewer receptors could lead to an impaired reward system. In short, the team theorized, people inheriting this gene might get less pleasure—less of an internal "high"—from enjoyable life events. When such people drink dopamine-boosting alcohol, they might feel "normal" for the first time in their lives. Such a feeling would be very powerful and could lead to the intense craving observed among alcoholics.

It was a beautiful theory, and it was fairly easy to understand. Not surprisingly, the story made the front page of the *New York Times* and many other papers. It was hailed in popular media as the ultimate confirmation of the disease theory of alcoholism and the humbling of the backward "nurturists," who claim that alcoholics are, to some extent anyway, responsible for their own condition.

Unfortunately, plausible though the D_2 theory is, it hasn't been embraced by the alcohol research community (Holden 1994). Six months after Blum and Noble's paper appeared, a team at the National Institute on Alcohol Abuse and Alcoholism announced that it could find *no* significant difference between alcoholics and nonalcoholics in the frequency of the A1 allele. In 1991 a group at Washington University in St. Louis also failed to find any association. And, more recently, a team at the National Institute of Mental Health, looking not at the allele but directly at the D_2 gene, also found no differences between alcoholics and nonalcoholics.*

None of these negative findings made the front page of any newspaper, if they were reported at all—a trend that is distressingly familiar to scientists in the field of genetics. The initial announcements of the "discoveries" of genes causing manic-depression, schizophrenia, and excessive violence among men with an extra Y chromosome all were trumpeted loudly in popular media. In each case, however, follow-up studies failed to confirm the initial findings, and in some cases the earlier reports were retracted (Horgan 1993). But, as happened in the case of the D_2 gene, none of these retractions and corrections received anything like the original coverage. This kind of uneven reporting has no doubt contributed to the popular assumption that genes are all-powerful and that alcoholism is purely a genetic disease.

Nature

Does all of this mean that the D_2 theory of alcoholism is dead? Not at all. Numerous studies continue to suggest that something is going on with dopamine in the brains of some alco-

---

*Blum and Noble were members of that team, and they disagree with their coauthors about the meaning of the results. They claim that the mutation in the gene possessed by alcoholics lies not in the part of the gene examined—the so-called exon sequences, which determine the *structure* of the dopamine receptor—but in some other section regulating the *number* of receptors made. To date, this hypothesized "intron mutation" has yet to be identified.

holics, particularly those with the most serious type of alcoholism. Evidence from twin studies, the use of dopamine-boosting drugs, and ongoing genetic screening studies indicate that the A1 allele is correlated to some extent with severe alcoholism. By one estimate, defects in the D_2 gene may account for about one-third of the overall influences on the prodigious use of addictive substances (Uhl et al. 1993). Other, unknown genes were estimated in this study to account for another third, and the last third was attributed to the environment. As the authors of the study noted, this model would leave D_2 mutations as "one of the most prominent single gene determinants of susceptibility to severe substance abuse—but other genes and the environment, when combined, still play the largest role."

That other genes, and other neurotransmitters, probably have a hand in alcoholism is hardly surprising given the brain's complexity. In fact, variations in several other neurotransmitter systems have been tentatively linked to alcoholism. One candidate is serotonin. The brains of alcohol-preferring rats, for instance, have been shown to have lower levels of serotonin than the brains of other rats (Murphy et al. 1987). And several animal and human experiments have shown that drugs *boosting* serotonin (such as Prozac) tend to modestly reduce drinking (Department of Health and Human Services 1993).

All these studies suggest a relationship between internal serotonin levels and the desire for alcohol. But none of the associations or effects observed are so significant that serotonin is being viewed as *the* alcohol-related neurotransmitter. For instance, one study of alcoholics given antidepressants that boost serotonin levels found that the drugs affected only about half of the study subjects, and that in this group there was only a 20 to 30 percent reduction in drinking (Naranjo et al. 1990). Such results may be explained by the hypothesis that serotonin is more responsible for regulating mood than for controlling alcohol cravings. In people suffering an underlying mood disorder, then, correcting serotonin levels could reduce their need for alcohol, while those whose disorder lies elsewhere derive no benefit.

Other research suggests that defects in the endorphin system of alcoholics may contribute to the condition. It has also been suggested that genetically based differences in alcohol metabolism may play a role. The bottom line is that the view of alcoholism as a single, clear-cut disease is considerably weaker and less substantiated than most people think. It's not that genes have nothing to do with alcoholism. We've just seen excellent evidence that they *are* involved. But the nature of that involvement and the degree to which it is manifested in a given alcoholic is still unknown.

For the moment, there is only the theory—as yet unproved—that genes set a kind of background tone for alcohol response. Many genes probably contribute to this tone. Some may influence how the body metabolizes alcohol; others may influence the sensitivity of GABA receptors; still others—such as the infamous D_2 receptor gene—may set the idle speed of the internal pleasure-producing machinery. Variations in the ways these genes work may leave some people more or less sensitive to alcohol than others. But no matter how much genetic variation is at work, genes don't function in a vacuum. They are expressed in human beings who live their lives in diverse and complicated environments.

Nurture

Researchers probing the environmental side of the alcoholism coin begin with the obvious question: Why do people drink? Instead of looking at the stage set by genes, these researchers look at what people *do*. They look at the choices people make in their drinking and at their behavior toward alcohol in general.

What they've found runs the gamut from the obvious to the surprising.

Many researchers were surprised, for example, by the results of studies on the relation of drinking to stress. As we've seen, alcohol mimics the actions of the antianxiety drug Valium, which might suggest that those experiencing marital, economic, or job-related stress would feel an increased urge to use alcohol. But study after study has found only small or negligible correlations between stress levels and drinking levels. Basically, the amount people drink has little to do with how much stress they're under.

Another environmental factor long thought to contribute to alcoholism is expectancy. Some people regard alcohol as a "magic elixir" capable of enhancing social skills, sexual pleasure, confidence, strength, and aggressiveness. When such beliefs have been acquired prior to the development of drinking problems, they have been associated with increased risk of alcoholism. This isn't surprising given what we learned in the previous chapter about how one's beliefs about the potency of alcohol change not only the subjective experience of consuming alcohol, but the physical responses as well.

One of the probable ways that many people *acquire* their expectations about alcohol is through advertising. The vigorous promotion of alcohol consumption in itself, however, apparently does not encourage people to drink. A before-and-after study of towns banning beer, wine, and liquor advertising found no subsequent change in total alcohol consumption (Smart 1988). And the lifting of advertising restrictions in Saskatchewan, Canada, had no overall effect on alcohol consumption, though drinking shifted slightly from spirits to beer (Makowsy and Whitehead 1991).

Another nongenetic component of alcoholism could be parental influence. Here, too, the evidence is surprising. Harburg and colleagues (1982) found that teenagers were more likely to *reject* than emulate parental behavior when parental drinking became extreme.

Although they are less influential than most people think, the kinds of environmental factors just mentioned probably interact in complicated and unpredictable ways with an individual's genetically set biological makeup. Accidents, viruses, severe abuse, emotional trauma, and learning can all change the brain's circuitry; purely environmental factors, in other words, can alter the biological foundations of behavior. And genetic factors— how we look, where our talents lie, and so forth—affect the way others treat us and the way we experience the world, and thus may exert considerable influence on the behaviors we develop. As one scientist put it, "Genes and environment loop out into each other and feed back on each other in a complex way that we have just begun to understand" (Mann 1994).

The Spice of Life

Human biological diversity is hardly a new concept. In the unsettled years during which Julius Caesar struggled to gain control of the Roman Republic, the poet Titus Lucretius Carus wrote a remarkable didactic poem called *De Rerum Natura* (On the Nature of Things). The work is a lavish ode to Lucretius's philosophical hero, the Greek philosopher Epicurus, who, among other things, anticipated by thousands of years such modern ideas as the atomic theory of matter, the universality of physical laws, and the molecular basis of individual differences. "With the outward difference between the various types of animal that take food," Lucretius wrote, "there go corresponding differences in the shapes of their component atoms. These in their turn entail differences in the chinks and channels—the pores, as we call them—in all parts of the body.

Substitute DNA for "component atoms" and "ion channels" for "chinks and channels," and you have a good approximation of today's view of the wellsprings of human individuality. It is now clear that if we could see a person's chromosomes as clearly as we can see her face, we would perceive in those long, spiral molecules the same degree of individuality.

If the acceptance of unique faces and bodies is ancient, however, the notion that *brains* are equally unique has taken longer to root. Perhaps it is a reflex belief that despite outward differences, inside we are all "created equal." But, of course, brains are no more the same from person to person than are fingerprints—each is an expression of a singular genetic heritage. Each of us has a unique number of neurons, neuronal connections, levels of neurotransmitters, and sensitivities in our ion channels, and thus unique responses to outside influences such as alcohol. The current struggles to pin down the mechanisms underlying both ordinary intoxication and alcoholism are driving this point home with a vengeance. The number of ways people vary in the details of their neural architecture, and specifically in their responses to alcohol, is astounding.

Some people are physiologically vulnerable to the ravages of alcoholism; others can take alcohol or leave it. Certain individuals get sleepy on low doses of alcohol and revved up on high doses—exactly the reverse of what most people experience. A person might get hangovers on white wine but not red, or require two hours rather than one to recover mental clarity after a single drink. Such variations can make hash of attempts to say anything categorical about how people respond to alcohol. Still, most of the effects discussed thus far, from the molecular to the behavioral, are true to some extent for all people. Understanding how alcohol *usually* works can provide a benchmark against which to measure one's own responses.

In our exploration of alcohol, we started with the perspective of a single molecule—a pudgy, dog-shaped assembly of nine atoms. By understanding the size, shape, and chemical properties of ethanol, we saw why it so easily soaks into the body and insinuates itself into the molecular machinery underlying functions as diverse as thought and urination. Then we pulled back a bit. We met other molecules as we followed a shot of scotch down the throat. Proteins. Ion channels. Enzymes that rip atoms off alcohol molecules. Understanding something about how these molecular machines work helps in understanding how alcohol itself works. After that, we pulled back a bit farther, to the size of cells. We met neurons, the fundamental units of consciousness, and saw how they generate action potentials, the "sparks" underlying all human behavior. Pulling back even farther we looked at the body as a whole— at how moderate doses of alcohol can help the heart, modify sexual response, or tweak the immune system. We also saw how heavy or long-term drinking can ruin these bodily systems and lead to impotence, enfeeblement, pain, or death. We then considered whole populations of drinkers: those for whom alcohol is addicting and those for whom it is not. We looked at how the nature of one's genes is just the flip side of the nurture of one's environment.

Now, pondering the foundations of our individuality, we are back at the level of molecules. We've seen how our unique DNA gives rise to unique brains, which in turn give rise to unique minds and personalities. The complexity of the human brain is the reason that alcohol is such a rich, complicated, exasperating subject. The molecule itself is laughably simple—as boring and static as a pinball. But let a few trillion of those pinballs fall into the machinery of the mind—into the flashing, deafening confabulation that is a human being—and anything can happen. Anything at all.

think I just saw Jesus in my cup of Taster's
ice.

—Zippy the Pinhead

The Eyelids of Bodhidharma
· · · · · · · · · · · · · · ·

6

The World's Favorite Drug

Alcohol is scarce in the natural world. Producing appreciable quantities demands somewhat laborious and delicate manipulations of yeast. Caffeine, in contrast, quite literally grows on trees. And bushes. And some kinds of cactus. And some species of lily and holly and camellia. In fact, at last count, more than a hundred plant species produce caffeine molecules in their seeds, leaves, bark, or other structures, making for a truly remarkable distribution (Viani 1993). Two other popular plant-produced molecules—nicotine and morphine—are roughly the same size and complexity as caffeine, but both are produced in only a single plant species: tobacco (*Nicotiana tabacum*) and opium poppies (*Papaver somniferum*), respectively.

Since caffeine-containing plants grow almost everywhere in the tropics, it is not surprising that the inhabitants of those regions long ago learned ways to extract the stimulating drug for their own uses. In Africa, caffeine was discovered in kola nuts and in the seeds and leaves of the many species of coffee tree, two of which are grown commercially: *Coffea arabica* and *Coffea robusta*. *Arabica* beans are harder to grow, produce more flavorful coffee, and contain about half the caffeine of *robusta* beans. In China, caffeine was discovered in the leaves of tea plants. And in South America, caffeine was found in the leaves of the maté plant (used to make a drink of the same name) as well as in the seeds or berries of several other plants used to make beverages no longer popular.

Tea and coffee have become the most popular drinks on earth. Aside from plain water, more tea is consumed every day by the world's people than any other single beverage (Graham 1984). Coffee is a close second, and because a typical cup of coffee contains about twice as much caffeine as a cup of tea, coffee is actually the single largest source of caffeine worldwide (Gilbert 1984). In the United States, soda, not tea, is the most popular beverage; per capita soft-drink consumption in 1993 was nearly 50 gallons. Coffee ranked second with 34 gallons consumed per person a year, and beer was third at about 23 gallons (Berry 1994; National Coffee Association 1991).

The popularity of soda in the United States hardly means that Americans prefer caffeine-free beverages. Roughly 86 percent of the 12.7 billion gallons of soda consumed in 1993 contained caffeine (Berry 1994). A good deal of this caffeine is found in cola drinks such as Coca-Cola® and Pepsi-Cola® and their many imitators. In fact, the word "cola" comes from "kola," the name of the African tree that produces the caffeine-containing seeds from which a flavor extract is made. This kola extract was one of the ingredients in the original recipe for Coca-Cola, invented by Georgia pharmacist John Pemberton in 1886. Pemberton's brew also contained cocaine, derived from the coca plant of South America, which is where the "coca" in Coca-Cola" comes from. After the addictive potential of cocaine was recognized around the turn of the century, the drug was eliminated from the recipe and replaced with caffeine. Today, both Coke and Pepsi contain about 45 milligrams of caffeine per 12-ounce can—roughly the same as a cup of tea or half a cup of coffee.

But non-cola soft drinks can contain significant amounts of caffeine as well. Mountain Dew®, for instance, contains 54 milligrams of caffeine per can. Mellow Yellow® and Dr. Pepper® also contain hefty doses. Interestingly, almost all the caffeine in these drinks is purchased by soda manufacturers from the makers of decaffeinated coffees and teas, for which caffeine is a valuable by-product indeed.

In America, therefore, soda consumption accounts for a significant percentage of total caffeine ingestion. Despite the phenomenal growth in an espresso-based café culture in many large cities, coffee consumption overall has been declining slowly over the past decade while soda consumption has risen steadily. If current trends continue, more Americans will get their caffeine buzz from soda by the turn of the century than from any other source.

A Brief History of Caffeine

The world's fondness for caffeine has been a long-standing love affair. Although the discovery and use of caffeine-containing plants predates writing, various legends and myths about the discovery of coffee and tea have survived to the present day.

The discovery of tea is attributed to the Chinese emperor Shen Nung. The year is fixed as 2737 B.C. According to legend, one evening the emperor was boiling water in an open kettle over a campfire built from the branches of a nearby shrub. Some scorched leaves from these branches swirled up in the column of hot air and fell back into the water. Rather than discarding the contaminated water, the emperor tasted it and was intrigued by the astringent taste and refreshing aroma. Further experimentation with more leaves of the same tree convinced Shen Nung of the value of the plant as a health-giving medicine. Over the centuries, the use of tea expanded from its initial role as a medicinal herb to that of a ubiquitous social beverage.

The custom of drinking tea was brought from China to Japan by Buddhist priests around the year A.D. 600 (McCoy and Walker 1991). This explains why the legend of tea's origin in Japan is linked to Buddhism, and in particular to Bodhidharma, the sage who founded the Zen branch of Buddhism. According to the legend, Bodhidharma fell asleep in the course of an extremely long meditation. Disgusted with his own weakness, he tore off his eyelids and flung them to the ground. Where the eyelids fell, tea plants sprang up, thus providing other Buddhist priests with a tool for extending the reach and power of their meditation.

The connection between caffeine and religious devotion also figures prominently in one of the common legends about coffee. In this myth, a sharp-eyed Arabian goatherd named Kaldi noticed his flock munching the bright red, cherrylike fruit of a shrub native to northeastern Africa. Soon after the goats ate the berries, they began prancing around with unusual gusto. Kaldi tried the berries himself and was so refreshed and invigorated that he danced along with his goats. This frolicsome behavior was noticed by a drowsy monk who was passing by on his way from Mecca. Impressed, the monk asked Kaldi the secret of his energy. Kaldi showed him the berries. The monk was delighted to find that he could now pray longer and with more attention. He spread the word to his fellow monks, who experimented with other ways to consume the berries. Eventually, people found that roasting the seeds, grinding them up, and soaking them in hot water produced a beverage that was tasty and gave a greater "kick" than could be achieved by merely chewing the caffeine-containing fruit and seeds.

Nature's Pesticide

As with the arts of fermentation and distillation, humans mastered the cultivation, processing, and preparation of caffeine-containing beverages long before they knew what gave these drinks their zip. But whereas alcohol was isolated from wine and beer by distillation in the Middle Ages, the active ingredient in coffee and tea remained mysterious until the nineteenth century.

The reason is that the separation of caffeine from tea and coffee is considerably more difficult than distillation. It requires several separate chemical steps and the use of a strong solvent, such as hexane or chloroform. It wasn't until 1820, at the dawn of the modern era of organic chemistry, that caffeine was discovered. The word "caffeine" comes from *café*, the French word for "coffee." The word "coffee," in turn, was coined by the great naturalist Carolus Linnaeus. He gave the name *Coffea* to the genus of tropical shrubs now called coffee trees. Linnaeus created his word as a Latinization of three of the many different words then used to describe coffee: "caova," "cova," and "kahwah."

By 1865 caffeine had been isolated not only from coffee, but from tea, maté, kola nuts, and various South American plants used to make aboriginal drinks. At a pharmaceutical meeting that year a Professor Bentley noted what many have commented on since. "It is most remarkable," he said, "that all the most important unfermented beverages in use in different parts of the globe should be prepared from substances containing the same or a closely allied alkaloid" (Kihlman 1977). It isn't clear whether Bentley was more impressed by the prodigious human thirst for stimulants or by the unusually wide distribution of caffeine among plants, but both facts are, indeed, remarkable. The alkaloid family referred to by Bentley is a large and generally poisonous group of nitrogen-containing compounds that includes strychnine, nicotine, morphine, mescaline, and emetine; the last of these is the deadly ingredient in poison hemlock, the herb used by the ancient Athenians to execute the philosopher Socrates.

As we've seen, most alkaloids are specific to a particular plant species. How is it, then, that caffeine is found in plants as unrelated as certain species of lilies and cacti? Why do more than a hundred species go to the trouble of manufacturing this one psychoactive molecule? One theory holds that caffeine helps plants ward off attack from insects and animals; caffeine may make seeds, leaves, and other plant parts taste bitter, thus discouraging consumption by predators. It is also possible that caffeine acts directly on animal nervous systems to create uncomfortable—or downright lethal—effects. For both these reasons, caffeine-containing plants may have a distinct adaptive edge over other plants, which would logically lead to their proliferation over other species.

The caffeine-as-defense theory is supported by research on caffeine's pesticidal properties. In the mid-1980s, James Nathanson of Harvard Medical School found that when prepara tions of tea leaves or coffee beans were fed to larvae of young tobacco hornworms, the insects ate poorly and suffered tremors, hyperactivity, and stunted growth (Nathanson 1984). At slightly higher concentrations, the larvae were killed within twenty-four hours. The same was true for mealworm larvae, a species of butterfly, mosquito larvae, and milkweed-bug nymphs. The effects were so pronounced that Nathanson suggested that caffeine be considered a viable natural alternative to stronger and more ecologically damaging pesticides.

Similar kinds of chemical self-defense systems have been discovered in many other plants, so the role of caffeine as a kind of natural nerve poison makes a good deal of sense. In fact, the same evolutionary principles may explain the existence of nicotine, morphine, cocaine, tetrahydrocannabinol (the active ingredient in marijuana), and other plant-derived psychoactive drugs. These compounds probably exist because they disrupt the nervous systems of insects and animals, thus conferring an adaptive advantage to plants that manufacture them.

This survival advantage is probably not, however, the entire explanation for caffeine's presence in so many kinds of plants. If it were, the growers of tea and coffee would have a much easier job. But despite the presence of caffeine-laced sap in the tissues of tea plants and coffee trees, many insects find both crops eminently palatable, forcing growers to rely heavily on pesticides. (In recent years, some growers have returned to organic approaches with some success, suggesting that the practice of growing dense monocultures of such crops as coffee and tea may contribute to their vulnerability to insect attack.)

The presence of caffeine in tea and coffee plants has, of course, provided these species with a serendipitous advantage: because caffeine is so highly prized by humans, the plants from which it is derived have been protected and distributed to an extent impossible under natural conditions.

The point is that caffeine's pesticidal properties, though real, are probably not the whole reason this compound appears in so many plants.

Caffeine: The Molecule

Another explanation for caffeine's wide distribution in the plant kingdom is that it's manufactured from a very common raw ingredient. The starting point for caffeine production is a ubiquitous molecule called xanthine:

Xanthine

Xanthine is found in both plants and animals. It is a raw material used constantly in the creation and maintenance of DNA. Xanthine isn't usually found in great quantities in animal bodies because it is rapidly recycled into other molecules by enzymes. In humans, for instance, excess xanthine is usually converted into uric acid, which—as the name suggests—is excreted in urine. In plants that make caffeine, however, xanthine is shunted into an enzymatic assembly line that methodically attaches common molecular units called methyl groups. A methyl group is a carbon with three hydrogen atoms attached.

When a methyl group is attached to xanthine, a *methyl*xanthine is formed. When a second is added, a *di*methylxanthine is formed. And when a third is attached, a *tri*methylxanthine is created. Caffeine is a particular kind of trimethylxanthine. Here is its molecular picture:

Caffeine

As you can see, caffeine is a globular molecule with a knobby surface. One consequence of this lumpiness is that caffeine is far more selective in its actions than is alcohol. Like a complicated key that can work only in an equally complicated lock, the unique bumps and grooves of caffeine mean that it interacts only with molecules that happen to mirror those shapes. This makes caffeine very picky in its actions—just the opposite of alcohol's tendency to affect practically every molecular system it touches. Caffeine, in fact, leaves the vast majority of the body's molecules alone. But it binds very strongly indeed to the select handful that happen to have the correct shape. In practical terms, this means that it takes much less caffeine than alcohol to achieve a desired physical or mental effect.

Unlike the standard alcoholic drink referred to earlier, there is no officially sanctioned "standard drink" of caffeine. The most commonly used unit is the 100 milligrams of caffeine found in an average (8-ounce) cup of regular coffee. Just as a shot of whiskey, a glass of wine, and a can of beer contain about a half-ounce of alcohol, an average cup of coffee, two cups of black tea, and two 12-ounce cans of caffeinated soda contain roughly 100 milligrams of caffeine. Notice that we're talking *milligrams* here. A milligram is a thousandth of a gram— roughly the weight of a single grain of salt. In contrast, the standard dose of alcohol is measured in liquid *ounces*. The half-ounce of alcohol in a standard drink is equivalent to 14,200 milligrams. The standard dose of alcohol is thus roughly 142 times larger than the standard dose of caffeine.

Actually, this comparison says more about the relative impotence of alcohol than the strength of caffeine. Caffeine's potency is similar to that of other familiar drugs like aspirin, which is typically administered in doses of between 50 and 500 milligrams. An example of a *truly* potent drug is lysergic acid diethylamide, or LSD. A typical "hit" of LSD is a mere tenth of a milligram. Picture cutting a single grain of salt into ten pieces and using just one of those pieces. That's how little LSD it takes to produce a neurological impact far more dramatic than that caused by a cup of coffee.

As different as they are in potency, however, caffeine and alcohol molecules have an important similarity: both dissolve easily in either water or fat. Just like alcohol, caffeine molecules are quickly absorbed in the small intestine (and, to a lesser extent, in the stomach itself). Caffeine easily crosses cell membranes, and is rapidly distributed to all of the body's tissues. It also diffuses into bodily liquids such as saliva, semen, breast milk, and amniotic fluid. Unlike alcohol, however, hardly any caffeine is eliminated in the breath or urine: it simply circulates until it is destroyed by enzymes in the liver.

Caffeine's Cousins

Caffeine isn't the only methylxanthine consumed by humans—it's just the most famous. Coffee and tea also contain very small amounts of a methylxanthine called theophylline.

Cacao products (that is, chocolate in all its guises) contain yet another methylxanthine called theobromine:

Theophylline   **Theobromine**

As you can see, both theophylline and theobromine are di-methylxanthines. Unlike caffeine, these molecules have only *two* methyl groups attached to their xanthine skeleton. You can also see that these compounds are extremely similar to each other: they contain exactly the same number and kinds of atoms. They differ only in the position of one of the methyl groups. In a beautiful example of how shape is everything in chemistry, this seemingly trivial difference produces striking differences in the way the two substances affect the brain. Theophylline is roughly as potent as caffeine; theobromine is seven times *weaker* than either.

Theophylline is perhaps best known for its medicinal qualities. Because it very effectively relaxes the bronchial passageways, it is often the drug of choice for treating asthma and other breathing difficulties, such as congestion caused by pet allergies. Indeed, allergy sufferers who are unaware of theophylline's similarity to caffeine may find themselves wide awake at night or unusually jittery if they take their medication and also drink their usual coffee, tea, or cola.

Caffeine opens up bronchial passages also—though less dramatically than theophylline—and this particular attribute is part of the story behind one of the most famous coffee slogans of all time.

Theodore Roosevelt was prone to asthma attacks when he was a boy. His doctor recommended small doses of coffee to arrest these attacks, which started Roosevelt on a habit that grew over time into a legendary appetite (Siegel 1989). His coffee mugs were said to more closely resemble vats than cups. In 1907 coffee merchant Joel Clark sought to take advantage of this prodigious thirst. Clark had set up a booth to display his wares at a county fair to which Roosevelt was paying a visit. When Roosevelt walked by his booth, Clark thrust a cup of his coffee at him—a cup the president promptly drained in a gulp. Setting down the empty cup, Roosevelt turned to the people around him and declared the coffee "good to the last drop," thus giving Clark and his brand of coffee—Maxwell House®— a slogan that lives to this day.

Caffeine's other cousin, the relatively weak theobromine, is consumed by countless millions every day. Most people know that chocolate contains a small amount of caffeine—roughly 20 milligrams in a 1-ounce portion. That's not much—only one-fifth the amount in an average cup of coffee. But most people *don't* realize that chocolate also contains theobromine. In fact, theobromine is *seven times* more abundant than caffeine in chocolate—about 130 milligrams in a 1-ounce piece. This abundance neatly compensates for theobromine's lack of raw pharmacological punch. Basically, when theobromine's influence is added to caffeine's, a 1-ounce piece of chocolate can be said to have the stimulating power of roughly 40 milligrams of caffeine, about the same as that found in a cup of tea (Gilbert 1992).

If it were only asthmatics who needed to appreciate theophylline, or "chocoholics" who needed to think about theobromine, these two methylxanthines would be mere curiosities to most other people. But, in fact, these two substances play an important role in the lives of everyone who drinks coffee, tea, colas, or any other caffeine-containing beverage.

The Long Buzz

Regardless of your fondness for a steaming mug of joe, a fragrant cup of tea, or an ice-cold Coke, your body responds to the caffeine in these beverages as though you had just swallowed poison. As with alcohol, liver enzymes are marshaled to attack the molecules and disable them as quickly as possible. The human liver disposes of caffeine by undoing the steps that led to its formation in plants: methyl groups are plucked off one at a time. This is an important point: depending on *which* methyl group is removed, caffeine is transformed into theophylline, theobromine, or another dimethylxanthine called paraxanthine.

As we just saw, theophylline is roughly as potent as caffeine, so when theophylline results from the first stage of caffeine metabolism the arousing effects of the original caffeine remain unchanged. Theobromine is only one-seventh as potent as caffeine, so the conversion of caffeine to this dimethylxanthine *does* represent progress. But 70 percent of a given dose of caffeine is converted to paraxanthine, which is actually slightly *more* potent than caffeine. This means that the "buzz" you get from a cup of coffee has as much to do with the breakdown products of caffeine as with the caffeine itself. Exactly how paraxanthine affects the brain is not clear, though it seems to mimic the actions of caffeine due to the similarity of its methyl-group configuration. (We'll delve into these actions in the next chapter.)

In the second step of caffeine metabolism in humans, another methyl group is removed, producing a methylxanthine, which has no stimulating effects. From there, the last methyl group is removed, yielding plain-old xanthine, which either is eliminated in urine or is reused. The pharmacological activity of theophylline, theobromine, and paraxanthine is part of the reason it takes a relatively long time for a coffee buzz to wear off. Not only must the caffeine be eliminated, but the break-down products have to be eliminated as well. The time it takes for a dose of a drug to wear off is measured by a value called a half-life. That's the time it takes for *half* of a dose to be eliminated. The half-life of caffeine averages between five and six hours, which is far slower than the rate at which we elim-inate alcohol. As leisurely as caffeine's half-life is, however, it can be even longer for certain people.

Women taking oral contraceptives require about *twice* the normal time to eliminate caffeine (Yesair 1984). For such women, the stimulation from a single cup of coffee might last all day. A similar, though less dramatic increase in caffeine's half-life has been reported for women during the luteal phase of the menstrual cycle—the time between ovulation and the beginning of menstruation. In one study, caffeine elimination took about 25 percent longer during this time, resulting in an average half-life of 6.8 hours (Arnaud 1993). And in infants, the half-life of caffeine is radically extended because their livers have not yet developed the enzymes needed to break down caffeine. A full-term newborn requires eighty hours to meta-bolize half a dose of caffeine (Snel 1993). As infants grow, their ability to process caffeine also grows. By the time a baby is between three and five months old, a dose of caffeine will have an average half-life of 14.4 hours. And by about six months, infants have essentially the same ability to process caffeine as adults. Although studies have failed to find any adverse con-sequences on infants from the caffeine consumption of nursing mothers, the extremely long half-lives in young babies is one reason that many doctors advise breast-feeding mothers to avoid caffeine altogether. (This applies to expectant mothers as well, as we'll see later.)

But another segment of the population experiences exactly the opposite effect as women on oral contraceptives. By a still imperfectly understood mechanism, cigarette smoking "revs up" the liver's caffeine-destroying enzymatic machinery (Benowitz et al. 1989). As a result, the half-life of caffeine among smokers is reduced to an average of three hours (Parsons and Neims 1978). This double-speed elimination of caffeine may explain the long-standing observation that smokers drink more coffee than nonsmokers. Smokers may simply be adjusting their caffeine intake to maintain the same degree of stimulation achieved by nonsmokers.

Interestingly, it is apparently *not* the nicotine in cigarette smoke that induces liver enzymes to work more efficiently. Cigarette smoke contains hundreds of other volatile, reactive compounds, and it is apparently a family of such compounds called polycyclic aromatic hydrocarbons that triggers the increased enzymatic activity.

The impact of smoking on caffeine clearance is important for those who quit smoking. In one study, blood-caffeine levels jumped an *average* of 250 percent a few days after the subjects had quit smoking—even though they didn't change their coffee- or tea-drinking habits. This added caffeine jolt could easily exacerbate the anxiety, insomnia, irritability, and other unpleasant symptoms of nicotine withdrawal experienced by quitters.

Smokers who drink coffee and other caffeine-containing drinks are juggling the pharmacological effects of two fairly powerful alkaloids: nicotine and caffeine. This juggling is mostly unconscious: they automatically adjust their consumption of both drugs to maintain a desired physical or mental state. But as many people know from experience, this juggling is tricky. Not only do variables such as food intake and sleep alter the body's response to both substances, but interactions such as the one just mentioned between cigarette smoke and caffeine metabolism can produce effects that can leave a user grappling with physical reactions that seem out of proportion to the amount of a drug consumed. Understanding the nature of caffeine and how it behaves in the body can inform the self-regulation efforts of smokers and nonsmokers alike.

Now that we know something about caffeine as a molecule, we're ready to take a look at what happens when those molecules hit an unsuspecting brain.

Who can tell to what extent coffee has affected the impetus of our modern civilization? What new physical comforts . . . , new explosives, and new progress towards peace have been propelled through the last ounce of mental effort generated by the extra thrust from coffee?

—Frederick L. Wellman, head of the Coffee Research Project, University of Puerto Rico

A Quicker Genius

The Taste of Pitch

When coffee was introduced to Europe by enterprising merchants in the middle of the seventeenth century, it was regarded as an exotic Arabian curiosity at best, a repulsive excuse for a beverage at worst. Three hundred years ago, many people couldn't imagine consuming a hot, bitter, black drink. It reminded them of hot pitch, which was used as a medieval weapon and instrument of torture (Schivelbusch 1992). But this initial aesthetic resistance to coffee quickly evaporated as the pharmacological powers of the drink became widely appreciated. Within a decade of its nearly simultaneous introduction to urban seaports such as London, Amsterdam, Venice, and Marseilles, coffee drinking was commonplace. Coffeehouses and cafés opened up by the thousands, and comments about coffee's effect on the mind began appearing in the press. "'Tis found already," wrote James Howell in 1660, "that this coffee drink hath caused a greater sobriety among the Nations. Whereas formerly apprentices and clerks used to take their morning's draught of ale, beer, or wine, which by the dizziness they cause in the brain made many unfit for business, they now play the good-fellows in this wakeful and civil drink."

The caffeine in coffee, tea, and chocolate—all of which were introduced to Europe at about the same time—neatly dovetailed with the ideals and values of Enlightenment society. "Coffee functioned as a historically significant drug," writes Wolfgang Schivelbusch (1992). "It spread through the body and achieved chemically and pharmacologically what rationalism and the Protestant ethic sought to fulfill spiritually and ideologically. With coffee the principle of rationality entered human physiology."

In short, people instantly recognized coffee for what it was: a potent stimulant that induced a mental and physical energy that was both pleasurable in its own right and eminently useful. Coffee's ability to foster industriousness was seized on by employers looking to boost productivity and clergy looking to reduce alcohol consumption among their flocks. An anonymous poem, published in 1674 during the early years of coffee's popularity, is typical of the kind of good press coffee often received (Schivelbusch 1992):

> When the sweet poison of the treacherous grape
> Had acted on the world a general rape;

Drowning our Reason and our souls
In such deep seas of large o'erflowing bowls,

When foggy Ale, levying up mighty trains
Of muddy vapours, had besieg'd our brains,
Then Heaven In Pity
First sent amongst us this All-healing berry.

Coffee arrives, that grave and wholesome Liquor,
That heals the stomach, makes the genius quicker,
Relieves the memory, revives the sad,
And cheers the Spirits, without making mad.

Coffee's popularity has thus always been intimately tied to its ability to powerfully alter the functioning of the human brain—to make "the genius quicker." As we saw in the previous chapter, the compound responsible for this "quickening" was isolated from coffee in 1820. But only in the past several years have neuroscientists made much progress in illuminating *how* caffeine revs up the brain.

Perchance, to Dream

The mechanics of caffeine came to be understood during investigations of one of its most obvious and well-known properties: the ability to temporarily banish sleep. The nineteenth-century homeopath Samuel Hahnemann recognized this clearly when he wrote, in 1803, that for coffee drinkers "sleepiness vanishes and an artificial sprightliness, a wakefulness wrested from Nature takes its place."

Nobody at that time had the slightest idea about how caffeine worked, but, interestingly enough, they did have what has turned out to be a good grasp of the fundamental process underlying wakefulness and sleep. It had long been suspected that people get drowsy because of a buildup of some sort of sleep-inducing chemical produced during the day as the brain "worked." This was the rather crude idea behind a pioneering experiment conducted by two French scientists in the early years of the twentieth century. René Legendre and Henri Pieron worked with pairs of dogs. One dog was allowed to sleep normally, while the other was kept awake for extended periods of time. (How the experimenters managed to keep their hapless subjects from falling asleep isn't related.) They then extracted from the sleep-deprived dogs a small amount of the fluid that continuously bathes both the brain and the spinal cord. When this fluid was injected into the brains of the well-rested dogs, they promptly fell into a long slumber, proving that a sleep-inducing substance is, indeed, present in the brains of tired animals.

Unfortunately for Legendre and Pieron, the tools then available for examining cerebrospinal fluid for minute amounts of organic compounds were wholly inadequate to the task of identifying the mysterious substance. The experimenters had found an important clue about the chemical underpinnings of sleep, but they never came any closer to understanding its true nature. Since that time, of course, sleep has been extensively probed, in both animals and humans. Thanks to volunteers who collectively have spent thousands of nights sleeping in laboratories with their heads wired like Christmas trees, we now know that sleep is far more complex than a simple, featureless slumber. Brain-wave patterns change dramatically throughout the night as we pass through four major phases of sleep; several distinct brain structures cooperate to induce and maintain sleep; and not one, but several substances have been put forth as candidates for the sleep-inducing chemical sought by the

French researchers at the turn of the century. Recent study of one of these substances, called adenosine, has revealed much about how caffeine works.

For a long time, adenosine wasn't given much attention. It was known primarily for its role in the transfer of energy in cells. As cells use energy, adenosine is produced as a by-product. The harder a cell works, the more adenosine is created. The excess adenosine is pumped outside the cell. But recent work has revealed that adenosine is much more than mere cellular exhaust. In yet another example of nature's parsimonious ways, the adenosine produced by working cells is used as a signaling molecule in an elegant, self-regulating control system. How this control system works to regulate sleep and wakefulness was described by Robert Greene and his colleagues at the Harvard Medical School (Rainnie et al. 1994).

During the day, neurons fire frequently as we go about our daily business. As they fire adenosine is produced and ends up floating around in the immediate vicinity of the neuron. Embedded in the membranes of neurons (and many other types of cells) are receptors designed specifically for adenosine. When adenosine latches onto one particular kind of adenosine receptor, a chemical chain reaction is triggered inside the cell that almost immediately opens ion channels in the membrane. The opening of these channels either directly inhibits a neuron from firing or reduces the amount of neurotransmitter released into the synapse. In either case, the net effect is to dampen activity of both the neurons producing adenosine and the neurons in the immediate vicinity. Adenosine thus functions as a kind of thermostat: it keeps neuronal activity within safe limits. If neurons fire excessively, adenosine builds up, which slows firing rates. Less firing, of course, means less adenosine being produced; hence adenosine inhibits its own production. If activity drops off too much, adenosine levels fall as well, releasing the "brake" on firing and allowing neuronal activity to rise again.

This self-regulating adenosine "thermostat" is typically very localized: only the cell producing the adenosine and a few neighbors are involved. But recently it has been learned that adenosine plays a crucial role in setting the overall arousal level of the brain. Two thumbtack-size patches of neurons located on the brain stem are particularly loaded with adenosine receptors. The neurons in these two areas fan out and touch nearly every other part of the brain. They are thus exceptionally powerful—stimulate them, and the activity of the entire brain increases; dampen their activity, and the entire brain "goes to sleep."

As we use our brains while we are awake, adenosine builds up in these areas and a "brake" is applied with ever-increasing force, gradually quieting activity all over the brain. We become drowsy and feel a keen urge to sleep. Once asleep, the adenosine outside the cells is reabsorbed and recycled for use in energy production the next day. As adenosine levels drop, the "brake" is released and we wake up.

Blocking the Brake

Adenosine resembles another of the brain's potent "brakes": GABA, which, as we saw earlier, is one of the neurochemical substances used by the brain to offset and balance equally powerful "accelerator" neurotransmitters such as glutamate. The brain thus resembles a car with several brake pedals and several accelerators. Interfere with any one of these pedals, and you'll affect the speed and action of the car.

Caffeine works by getting in the way of the adenosine brake.

The reason becomes clear when the two molecules are compared:

Caffeine

Adenosine

Although at first glance these molecules may look dissimilar, close inspection reveals that each is built on an identical double-ring of carbon and nitrogen atoms. The primary difference between the two molecules is the addition of a type of sugar on the lower left side of the adenosine molecule. Otherwise, the two molecules are really quite similar indeed—to the point that caffeine easily masquerades as adenosine in the brain. In fact, because it's slightly smaller than adenosine, caffeine fits *more* snugly into adenosine receptors than does adenosine itself. This tighter fit enables caffeine to plug the receptor, thus preventing adenosine from binding.

Despite this aggressive attraction to the adenosine receptor, caffeine doesn't actually make a perfect fit. And this makes all the difference. If caffeine mimicked adenosine exactly it would be a depressant, not a stimulant: it would simply exacerbate and extend adenosine's natural inhibition in the brain. But when caffeine binds to the adenosine receptor, the resulting shift in the shape of the receptor molecule is slightly different from the warping that occurs when adenosine itself binds. As a result, the chemical chain reaction normally initiated by adenosine *isn't triggered by caffeine*. Caffeine is an impotent impostor: it binds with gusto, but fails to launch the all-important quieting message delivered by adenosine. Drinking caffeine is thus like putting a block of wood under one of the brain's primary brake pedals. Caffeine is an *indirect* stimulant: brain activity speeds up because it can't slow down. By itself, then, caffeine can't stimulate anything. It can only clear the way for the brain's own stimulants—neurotransmitters such as glutamate, dopamine, and the endorphins—to do their job. You can, therefore, get wired only to the extent that your natural excitatory neurotransmitters support it.

Among other things, this explains why it's almost impossible to overdose on caffeine. Even if every adenosine receptor in your brain were blocked by caffeine, you could still function. You would certainly feel stimulated, since one of your brain's main "brakes" would be disabled. But other brakes, such as GABA, would still be functioning and in the absence of any extra direct stimulants overall activity wouldn't kindle into the kind of neural conflagration that can occur with overdoses of drugs like cocaine and amphetamine.

Caffeine's relative safety is such that the only known deaths attributed to caffeine overdoses have been accidents. The lowest dose of caffeine known to have killed an adult is 3,200 milligrams, injected into a patient by a nurse who thought the syringe contained another drug (Gilbert 1992). A fatal dose of caffeine taken by mouth would have to be at least 5,000 milligrams: the amount in about forty cups of strong coffee consumed very quickly. Since this quantity of coffee would induce vomiting long before the lethal dose was reached, it's not likely that even the most determined person could commit suicide with coffee. In all cases of accidental death by caffeine, the actual cause is not an overstimulation of the brain but cardiac arrest induced by uncontrolled firing of the nerves that regulate heartbeat.

Another consequence of caffeine's indirect action in the brain is that ingesting more of it doesn't necessarily result in greater stimulation. If the levels of your excitatory neurotransmitters are low, no amount of caffeine will boost them up, and blocking more adenosine receptors with caffeine will have little effect. But there is another factor at work here as well. Animal experiments have demonstrated that caffeine is biphasic, which means that it has one effect at low doses and other effects at high doses. In one typical study using mice, caffeine stimulated activity up to doses roughly equivalent to three to four cups of strong coffee. As the dosage increased above that level, activity dropped. By the time dosage rose to the level equivalent to the caffeine content of ten cups of coffee, activity was lower than it had been before the mice had ingested *any* caffeine. In other words, in mice at least, high doses of caffeine act as a depressant (Seale et al. 1988). How mice actually *feel* after consuming large amounts of caffeine is, of course, unknown. But the results are intriguing because they seem to indicate that caffeine is affecting more than just adenosine receptors in the brain.

The current theory is that high doses of caffeine have a depressant effect because the caffeine interferes with another molecular regulator in the brain—an enzyme called phosphodiesterase. By blocking the activity of this enzyme, very high levels of caffeine may set off a chemical chain of events that inhibits neuronal firing.

The point of all this is that caffeine doesn't work the way most people think it works. The correlation between increasing dose and increasing stimulation applies only at doses equivalent to the caffeine content of between one and four cups of coffee. Beyond that, pouring more caffeine into the brain probably won't increase stimulation—and it may have the reverse effect because of caffeine's actions on other molecular subsystems.

These new findings about caffeine explain some of the ordinary experiences of drinking coffee, tea, and other caffeinated beverages. But the whole story is still not in hand. For one thing, adenosine isn't the only neurotransmitter involved in sleep and waking. Sleep seems to restore many brain circuits and systems, not just those using adenosine. Delaying or reducing sleep by using caffeine may, therefore, have subtle effects on mood or cognition that nobody has yet explored.

The Think Drink?

The common experience of having one's brain "turned on" by caffeine has been vividly described by Nobel Prize–winning scientist Leon Cooper, a professor at Brown University who is deeply involved in probing the brain. "This happens to me every morning of my life," he told writer George Johnson. "I just sit there at home with my *New York Times* and my big pot of tea, and after I have enough caffeine in me I can just feel my brain going from a barely conscious level to this high pitch, as though I've taken a drug. I'm suddenly enormously awake and very manic, as you can see. Ideas tumble out—almost all to be discarded by noon, unfortunately. But if I can focus on something where I really know the facts, where I really understand the problem, that's when something might happen" (Johnson 1991).

Certainly, Cooper isn't alone in his experience. Through the centuries, millions of people have used caffeine to help spark creative thought. One of the more famous is Johann Sebastian Bach. Bach passionately loved coffee, and in some ways no music better captures the essence of the caffeinated experience than some of his more frenetic fugues. Bach went so far as to write a musical ode to his favorite drink, the *Coffee Cantata*, a humorous one-act operetta about a stern father's attempt to control his daughter's fondness for the bean. "Dear father," the girl implores at one point, "Don't be so strict! If I can't have my little demitasse of coffee three times a day, I'm just like a dried-up piece of roast goat!"

Honoré de Balzac, the great French writer, also loved coffee and used it heavily. He typically went to bed at 6:00 P.M., arose at midnight, and wrote for twelve-hour stretches, drinking coffee continually. "Coffee falls into your stomach and straightaway there is a general commotion," he wrote. "Ideas begin to move like the battalions of the Grand Army on the battlefield ... things remembered arrive at full gallop."

Philosophers, too, have turned to caffeine to fuel their ruminations. Immanuel Kant, Jean-Jacques Rousseau, and Voltaire all adored coffee. The Scottish philosopher James MacKintosh even quipped that "the powers of a man's mind are directly proportional to the quantity of coffee he drinks."

As interesting as such examples are, of course, they in no way prove that caffeine actually improves mental functioning. It could be that Bach, Balzac, and Voltaire would have been just as creative without caffeine. Even after decades of investigation into caffeine's power to improve mental and intellectual function, it is still far from clear that coffee really is the "think drink." No one doubts that caffeine raises the overall arousal level of the brain, delays the onset of sleep, and heightens alertness. But whether these effects translate into clearer thought, better writing, or more creativity is an open question.

At the molecular level, the evidence is modestly encouraging. For instance, caffeine and other methylxanthines have been shown to enhance long-term potentiation—the enduring synaptic changes that are thought to underlie memory formation

(Tanaka 1990). Neurons bathed in modest levels of caffeine respond more vigorously to stimulation and form longer-lasting changes in their connections with other neurons.

Another line of research is also encouraging. Work from several laboratories has suggested that emotional arousal plays a critical role in memory. Basically, our strongest memories are of things that are emotionally provoking, in either pleasant or unpleasant ways. When adrenaline, the classic "fight-or-flight" hormone, is released during such arousal, it seems to "prime" the brain to remember things in unusual clarity. It is theorized that this mechanism evolved as nature's way of ensuring that animals clearly remember dangerous or provocative situations they encounter. This adrenaline connection to memory may explain the often-noted phenomenon of people remembering exactly what they were doing when they heard shocking news, such as the fact that John F. Kennedy had been shot or that the space shuttle *Challenger* had exploded.

It is at least theoretically possible that by stimulating emotional arousal and, specifically, by increasing levels of adrenaline, caffeine may prime the brain the same way that provocative experiences do. In one study, a 250-milligram dose of caffeine raised adrenaline levels 207 percent and noradrenaline levels 75 percent (Garcia 1993). The precise mechanism behind this apparent adrenaline boost remains unexplained.

Somewhat surprisingly, the effect of caffeine on memory formation has not been well studied in humans, though several experiments with animals have yielded positive results. In one such study, rats given caffeine before learning their way through a maze showed improved performance in later trials with the maze than did rats given a placebo (Battig and Wetzl 1993).

As suggestive as such results are, their meaning becomes elusive when considered in light of the many studies of performance conducted with humans. Here the data are often contradictory and difficult to interpret. In general, the studies have shown that caffeine improves mental ability on tasks requiring "speed," but degrades or has no effect on it with tasks requiring "power." For instance, caffeine is helpful in relatively passive, automatic, "data-driven" tasks such as auditory reaction time, visual-choice reaction time, and performing simple arithmetic. It also improves the ability to attend to something in a focused, sustained way. But in studies of more complicated tasks such as logical reasoning, numerical reasoning, reading comprehension, and complicated arithmetic—all of which require greater "central processing power"—caffeine either has had no detectable effect or has actually degraded performance.

Such findings suggest that famous coffee drinkers such as Bach and Kant may have derived little help from their caffeine habits. And yet, despite scores of studies on the subject, it is still too early to make such a judgment. For one thing, the tasks studied to date hardly capture the range of activities engaged in by humans; there is a great deal of difference between numerical reasoning and philosophical synthesis. For another, very few studies have taken into account the huge variation in caffeine response exhibited by individual human beings. In one study in which these differences *were* considered, the researchers found indications that caffeine's effect on mental performance varies with the impulsiveness of the user (Gilbert 1992). Impulsive people, in this study, were defined as those who tended to sacrifice accuracy for speed in various tasks and who tended to be more aroused in the evening than in the morning. When such people were given caffeine in the morning (when they were normally subdued), their performance at tasks such as proofreading for grammatical and typographical errors improved. But when caffeine was taken in the evening, their performance was *worse* than when they had no caffeine at all.

Conversely, people who scored low on impulsivity tests reacted in exactly the opposite manner, scoring worse in the morning and better in the evening.

In short, there is no simple answer to whether caffeine is, or is not, helpful in performing intellectual tasks. The answer appears to depend both on the nature of the task being considered and on the nature of the person doing the task. The general rule seems to be that caffeine is helpful for people who are not already naturally aroused and who are working on relatively straightforward tasks that don't require a lot of subtle or abstract thinking. But even this seemingly logical conclusion must be taken with a grain of salt. Is writing a straightforward task? Is musical composition? Is computer programming? Perhaps not for many people, and for folks who find such activities difficult caffeine may, indeed, be more of a hindrance than a help. But maybe for other people such "complex" tasks are more akin to play. Perhaps our notions of which tasks are straightforward and which are complicated are too limited. It may well be that the arousal induced by caffeine produces results that are as unique and difficult to predict as the actions of any given individual.

If this is the case, then even the most rigorous research in this area will be a poor guide for any particular caffeine user. People trying to determine whether caffeine helps or hurts performance in a given sphere of life will be left with no alternative but to become their own scientists and do their own research. They will simply have to experiment with their favorite source of caffeine, trying different doses at different times in an effort to see what works for their own unique biochemistry.

Coffee makes a sad man cheerful; a languorous man active; a cold man warm; a warm man glowing; a debilitated man strong. It intoxicates without inviting the police.

—John Ernest McCann, Over the Black Coffee

The Body, Wired

8

A Physical Drink

Early coffee and tea merchants touted their wares as much for the beverage's therapeutic effects on the body as for their stimulating effects on the brain. One 1657 advertisement in a London newspaper boasted that coffee was "a very wholesome and physical drink that helpeth digestion, quickeneth the spirits, maketh the heart lightsom, is good against eye-sores, coughs, colds, rhumes, consumption, headache, dropsie, gout, and scurvy." Such promotional enthusiasm was seldom encumbered by logic: coffee was said to both whet and curb the appetite; to increase alertness, but also to induce sleep; to cool "hot" temperaments while simultaneously warming "cool" temperaments.

No one today views coffee or any other caffeine-containing drink as this kind of medicinal panacea. But without a doubt, caffeine affects the rest of the body as powerfully as it affects the brain. It alters the functioning of nearly every bodily system, from the blood vessels in the head to the muscles controlling the toes and everything in between. For most people, these side effects are minor annoyances and are irrelevant to the desired effect of mental stimulation. But caffeine's side effects are actually the main point for others. Many athletes, for instance, take caffeine to boost their physical performance, and there are people who consume caffeine to curb their appetite, to lose weight, or as an inexpensive laxative.

Caffeine owes its remarkably broad reach to two related facts. First of all, the brain controls many bodily functions, either directly via nerve impulses or indirectly via hormones. By altering the brain, caffeine automatically alters all systems regulated by the brain. Caffeine also directly affects many parts of the body by attaching to adenosine receptors found outside the brain. Here, as in the brain, caffeine sometimes causes an accelerating or tensing response. It causes the heart to beat more rapidly, it constricts some blood vessels, and it causes certain types of muscles to contract more easily. But, paradoxically, caffeine can also cause a relaxation response. It can, for instance, relax the airways of the lungs and cause certain types of blood vessels to open.

These contradictory effects can be explained by probing deeper into the subject of adenosine receptors. To this point, we've dealt with only one type of adenosine receptor—the kind in the brain that cause neurons to slow down. But there are at least *three* types of adenosine receptors, all of which differ in the signals they send after adenosine binds to them. A_1 receptors, the kind we've been dealing with thus far, quiet activity by inhibiting cell firing. But A_2 receptors set off a different biochemical cascade when adenosine binds. This cascade sends just the opposite message: it excites neurons to fire. The chemical signals sent by A_3 receptors, the most recently identified, are not yet understood (Salvatore 1993).

For our purposes, it is not particularly important to know which kind of receptor is responsible for which bodily reaction, though the issue is of great interest to drug makers who hope to find molecules that bind selectively to just one type of receptor. Here it's enough to understand that adenosine receptors come in several varieties and that this is the fundamental reason that adenosine—and thus caffeine—have such apparently contradictory effects in different parts of the body. This chapter will examine some of caffeine's most interesting repercussions, from the promotion of urination to the elimination of headaches.

The Enfeebling Liquor

Unlike alcohol, caffeine is seldom viewed as an aphrodisiac. More typically, coffee, tea, and other caffeinated drinks are associated with work—particularly work involving thinking, reading, writing, or talking. Although such intellectual activities might for some people constitute sexual foreplay, this isn't their usual role and caffeine is not generally associated with sex in popular culture. Quite the opposite. Since its introduction in Europe, caffeine has more often been linked to celibacy at best, impotence at worst. The two most popular legends about the origins of coffee and tea involve using these substances to achieve a greater communion with God—hardly a ribald activity. The followers of Bodhidharma, remember, used tea to enhance meditation, and the monks of Mecca used Kaldi's discovery of coffee to extend their ability to pray.

In 1764, a broadside appeared on the streets of London: "The Women's Petition against Coffee, Representing to Publick Consideration the Grand Inconveniences accruing to their SEX from the Excessive use of that Drying, Enfeebling LIQUOR" (Schivelbusch 1992). The authors of the tract—a group calling itself the Keepers of the Liberty of Venus—were afraid that coffee would make "men [as] unfruitful as those deserts whence that unhappy berry is said to be brought."

Despite this apparent concern for the sexual health of men, the petitioners may have had an ulterior motive. Coffeehouses at that time were springing up by the thousands, and they were usually men-only establishments. The women supporting the petition may have been aiming a clever attack on discriminatory coffeehouses by hitting at a classic male weak spot. Nonetheless, variations on the "enfeebling" theme continued to be played. The nineteenth-century poet-historian Jules Michelet described coffee as "antierotic." "Coffee," he said, "at last replaces sexual arousal with stimulation of the intellect." Michelet's view was hardly idiosyncratic. At that time, coffee was often recommended as the preferred drink of clerics and other celibates.

Caffeine received no better press in the twentieth century. In a 1931 book about narcotics and stimulants, L. Lewin wrote that coffee could "sterilize nature and extinguish carnal desires." Today nobody views coffee as "sterilizing." But neither does coffee or any other caffeine-containing beverage connote sexual virility or physical passion. The romance and mystique surrounding coffee and tea remain primarily intellectual. If passion is involved, it is a passion for ideas and for their lively exchange under the influence of a drug that encourages quick thinking and wit.

It would be satisfying, at this point, if modern science could be called on to offer evidence that would either substantiate or reject these centuries-old views. Unfortunately, such data do not exist. Unlike alcohol's effects on human sexual response, which has been the focus of hundreds of papers over the years, the literature on caffeine and sex is anemic to say the least. If the subject is mentioned at all in scholarly journals or books, it is simply to note, as one author did, that "no specific effect of caffeine on sexual function has been demonstrated" (Hollister 1975).

The only aspect of sex that has been examined in any detail with respect to caffeine is rather far removed from the act itself. Several studies have demonstrated that sperm exposed to caffeine swim faster and more energetically than normal. In short, they wiggle more, allowing them to penetrate cervical mucus more readily (James 1991). The upshot is an increased chance of fertilization if an egg is nearby. These findings led some researchers to seriously consider the use of caffeine for human in vitro–fertilization therapy, since one of the big problems with sperm stored at extremely low temperatures is that when thawed they have "lowered motility." Unfortunately, however, the caffeine concentrations required to kick-start sperm in a glass dish are more than a thousand times higher than the level of caffeine in the body following even large doses of coffee. In fact, the 1,500 milligrams per liter concentrations used in some experiments may induce chromosomal damage, making this method of enhancing fertility dubious indeed.

The effect of caffeine on sperm motility is the only aspect of human sexuality that appears to have been examined in any rigorous way. One possible explanation for this is that caffeine's effect on sex is so trivial that nobody has felt motivated to go to the trouble and expense of measuring it. Indeed, caffeine's relationship to sexual response may be so slight that it's swamped by the much stronger variable we met earlier: the mind. We saw that people's expectations about alcohol can override the significant physiological effects exerted by that drug. It is likely that when it comes to caffeine and sex, people feel what they expect to feel: expect caffeine to be an aphrodisiac, like alcohol, and it probably will be; expect it to snuff out your carnal desires, and it will probably do just that.

Legal Speed

In December 1993, things looked good for twenty-four-year-old Sylvia Gerasch, the European champion in the 100-meter breast stroke. But only a month later she was stripped of her swimming title, dropped from the German racing team, and banned from further competition for two years (Reuters 1994). The reason? She drank too much coffee.

In January 1994, as part of the routine drug testing of athletes, Gerasch was found to have 16 micrograms of caffeine per milliliter of blood—a level considerably higher than the official limit of 12 micrograms per milliliter. Officials of the European Swimming Federation pointed out that such levels could be achieved only if Gerasch had consumed the equivalent of about eight cups of strong coffee in a short period prior to the testing. Gerasch protested the ruling, saying that she was only drinking her normal amount. But she didn't say what that amount was, and the officials stood firm.

Gerasch is hardly an isolated case. Other swimmers, runners, and bicycle racers have been penalized in recent years for excessive use of caffeine—a practice that for many years has been widespread but not always regulated. The issue first came to a head in 1962, when the International Olympic Committee listed caffeine as a "doping agent"—that is, a substance (such as steroids) that provides an artificial performance boost. In a controversial move, the committee removed caffeine from the list in 1972, but then reinstated it in 1984. Although this action surprises people who don't consider caffeine a drug—much less a "doping agent"—current research strongly supports the Olympic Committee's actions.

Caffeine has long been used to increase physical endurance, in both humans and animals. In Tibet, horses and mules working at extremely high elevations are often given large vessels of tea to increase their capacity for work (Gilbert 1992). The animals' masters, too, keep themselves going with caffeine. The distances between Tibetan villages is sometimes reckoned by the number of cups of tea necessary to sustain a person traveling that route, three cups of tea being roughly equal to 8 kilometers.

Studies of caffeine's effects on athletic performance have, for the most part, corroborated such anecdotal observations. For instance, recent research at Christ Church College in Canterbury, England, showed that caffeine reliably increased performance in several running events. Eighteen runners of various abilities ran a 1,500-meter time trial on two occasions: first after drinking two cups of coffee containing a total of 300 to 350 milligrams of caffeine, and then after drinking the same amount of decaffeinated coffee. Fourteen of the eighteen subjects ran faster after ingesting the caffeine. On average, their times improved by four seconds. Another test measured caffeine's effect on "burst" activity: the ability of ten competitive runners to "kick" the final 400 meters of a 1,500-meter time trial. The runners covered the first 1,100 meters at a predetermined pace and were then instructed to run the final leg as fast as they could. All ten "kicked" faster on caffeine—and also produced less fatigue-inducing lactic acid in their blood.

In another study, bicyclists who took 330 milligrams of caffeine one hour before exercising were able to pedal an average of 19.5 percent longer than subjects who drank no caffeine (Burke 1992) Impressed by results such as these, many professional bicyclists drink caffeine-containing beverages both before and during a race. Some even insert caffeine suppositories before a race in an attempt to provide a sustained dose with no stomach upset.

Exactly how caffeine boosts physical performance is still not clear. Attention originally focused on the way caffeine affects muscle contraction and reflexes. In 1913, the scientist Storm van Leeuwen reported that caffeine increased spinal reflexes in cats. Contemporaries reported similar increases in the familiar "kneecap," or patellar, reflex in humans. But these and other early studies could not be replicated later (Battig and Wetzl 1993). More recent studies of individual muscle fibers isolated outside the body show that caffeine can increase the speed and force of contraction, but only when caffeine levels are significantly higher than those found in the blood of even the most caffeine-abusing athlete (Fredholm 1984). A relatively new line of research suggests that muscle contractions may be stimulated by dopamine released by caffeine in the brain (Josselyn and Beninger 1991). But this theory is still quite sketchy and tentative. In the meantime, attention among sports physicians has turned away from the muscles themselves and toward the fuel that powers muscular activity in general.

Many studies have found that caffeine releases fat stored in cells and breaks it down into the smaller fatty-acid chains that the body burns as fuel. Caffeine may release these fatty acids directly via some as-yet-unknown process, or it may do so indirectly by raising adrenaline levels. Regardless of the mechanism, caffeine's ability to liberate some of the fuel supply stored in fat may explain its beneficial effects on athletic performance.

This effect can be strongly influenced by diet. One study

found that the greatest release of fatty acids occurred after athletes ate a rather atypical meal of sausage, bacon, and eggs (Weir et al. 1987). The effect was negligible when athletes ate a more common "carbohydrate loading" breakfast of cereal, toast, and orange juice. It has also been found that peak free-fatty-acid levels occur three to four hours after caffeine ingestion. This suggests that caffeine may be acting on more than just fat liberation, since it has been shown to boost performance in both short-duration and endurance sports. It may be that athletes engaging in short-duration events, for instance, benefit from caffeine's stimulation of the central nervous system, while endurance athletes may get their boost from the longer-term increase in fatty acids released by caffeine after several hours.

Whether caffeine helps or hurts a given athlete is a matter of individual response. Some athletes find it helpful, while others find that they suffer from an acid stomach, increased nervousness, or dehydration and thus avoid it. In any case, the current legal limits for caffeine are sufficiently high to allow a wide latitude of experimentation. The Olympic legal limit of 12 micrograms per milliliter (which has been widely adopted for other sporting events) is equivalent to about six cups of coffee consumed within thirty minutes. Since the results obtained in the studies mentioned earlier were found at levels considerably below this limit, it appears that caffeine is likely to continue to be one of the most widely used doping agents in athletic competitions.

Liquid Diet

The active ingredient in most over-the-counter diet aids used to be nothing more than a 200-milligram dose of caffeine— the equivalent of about two cups of coffee. Such pills were a gold mine for drug companies because caffeine is a relatively inexpensive raw material and customers are willing to pay handsomely for anything promising easy weight loss. But in 1991, the Food and Drug Administration banned caffeine from all weight-reduction products, ruling that caffeine neither suppresses appetite nor directly causes weight loss. (The companies selling the pills promptly replaced caffeine with another stimulant: phenylpropanolamine, one of the mildest members of the large amphetamine family.) Despite the FDA ruling about caffeine, however, many people still believe that a cup of joe will help them lose weight. Interestingly enough, caffeine *does* have some demonstrable effects that, at least theoretically, could help dieters. The problem—as the FDA recognized—is that for most people these effects are minimal to the point of insignificance.

We just saw one effect that might seem to justify the dieter's belief in caffeine: its ability to release fat and break it down into useful fatty acids. This is potentially significant for athletes because they are exercising so hard that their muscles readily burn the liberated fatty acids. But for more sedentary types, the fatty acids released by a cup or two of coffee are likely to simply be reconverted to fat once caffeine levels drop. Caffeine, in other words, isn't a "fat burner," but a "fat releaser." Exercise is still needed to effect weight reduction. Caffeine by itself simply sets the stage—and even then the effect is relatively small and of use primarily to athletes for whom even a 2 percent increase in energy availability might prove to be the winning margin.

What about another factor sometimes touted as being significant to dieters: caffeine's ability to rev up the body's metabolism? Several studies have found that moderate consumption of caffeine *does* raise the basal metabolic rate—probably by slightly increasing adrenaline levels—and that this results in a small elevation of body temperature and corresponding increase in caloric consumption. This rise, however, is very small indeed. Over the course of a day, the average increase in calorie consumption is between fifty and a hundred calories (Battig and Wetzl 1993). Given a perfectly uniform diet, this could have an effect since even a small increase adds up over the long haul. But in reality, people's diets are far from uniform and readily respond to variables—not the least of which is hunger itself—that are far more powerful than the modest temperature-raising effects of caffeine.

What about the possibility that caffeine curbs hunger by interfering with the brain's appetite-control center? Again, there may be something to this idea, but the necessary studies on humans haven't been done. It is theoretically possible that caffeine somehow inhibits parts of the brain, such as the hypothalamus, involved in regulating appetite. But nobody has yet suggested a plausible mechanism for such inhibition, and animal studies have found no relation between caffeine and food intake. It's unlikely, therefore, that any of these caffeine side effects plays a significant role in weight loss. In fact, for some people, caffeine may actually make it *harder* to eat a balanced, healthy diet. A study by researchers at the University of Michigan found that of 171 patients with eating disorders, those who consumed a lot of caffeine (more than 750 milligrams, or the equivalent of about eight cups of coffee a day) binged, fasted, and used laxatives and diet pills more often, and were more likely to smoke and abuse alcohol, than patients whose intake was moderate (Livermore 1991).

This is not the same as saying caffeine *causes* increased binge eating—just that there is an association between the two. It could well be that whatever underlies the pattern of extreme dieting also supports high use of caffeine. But the finding has renewed interest in the relationship between caffeine and ingestive behaviors and is of interest to those studying eating disorders.

Tremors and Twitches

Caffeine's ability to increase repetitive or involuntary muscle movements has received little scientific attention. A common example of such an effect is "treadle foot"—the rapid jiggling or pumping of one or both legs, sometimes observed in people who have consumed caffeine and are sitting down. Another well-known effect of caffeine is increased hand tremor, which has been measured in numerous experiments. And some people find that caffeine increases the number and intensity of small muscle twitches in such places as the eyelids, arms, or legs.

The mechanism behind these phenomena is still not well understood. The muscles involved in such tremors and twitches are skeletal, as opposed to cardiac or smooth muscle. Some experiments on skeletal-muscle strips have demonstrated that caffeine increases contractions, which might seem to explain some of the twitch phenomena. But the concentrations of caffeine required to produce such contractions are almost a hundred times higher than levels found in people's blood after ingestion of moderate amounts of caffeine.

Another possibility is that caffeine affects skeletal muscles indirectly. Experiments on frogs show that caffeine can release acetylcholine, the neurotransmitter responsible for initiating muscle contraction.

Regardless of the causative agent, the tremors and twitches experienced by some users of caffeine are usually harmless. Still, some heavy drinkers of caffeinated products experience cardiac arrhythmias (irregular heartbeats) or palpitations (a fluttering heartbeat), both of which, while not usually lethal, could be problematic for people with chronic heart conditions.

Prompting Nature's Call

One of the most well-known (if least-discussed) side effects of caffeine is its stimulation of urination and defecation. Like alcohol, caffeine increases urination both directly and indirectly. Since caffeine is usually consumed in beverages, the liquid by itself will result in an increased urge to urinate. But the kidneys are also rich in adenosine receptors. Adenosine helps regulate the delicate balance between blood flow and urine output. When caffeine blocks these receptors, blood vessels dilate, increasing the filtration rate and producing more urine. Both caffeine and alcohol thus are mild diuretics—drugs that increase urination—though the two substances achieve this end by different means. Caffeine's diuretic effect is usually mild and harmless. But for athletes and others who are likely to perspire heavily, excessive consumption of caffeine could lead to dehydration.

Caffeine is also a laxative. Like the kidneys, the colon is well endowed with adenosine receptors. Here adenosine helps control the tone of the smooth muscles used to propel feces on its way. Here, too, adenosine is the signaling molecule used to maintain the balance between relaxation and contraction. But unlike the action in the kidney, where caffeine causes a dilation, in the colon it causes a *constriction*. When adenosine receptors in the colon are blocked by caffeine or other methylxanthines, the normal relaxation messages are blocked. The smooth muscles thus more easily contract in a characteristic rhythm called intestinal peristalsis. Even moderate doses of caffeine can set off this peristalsis whether or not the body was ready to dispose of its feces. Although in some cases caffeine can cause diarrhea—typically among those who seldom consume caffeinated beverages—no harm from this laxative effect has ever been reported.

Of Headaches and Painkillers

Headaches afflict millions of people every day. The vast majority—about 90 percent—are caused by excessive tension in the head and neck muscles. But about 8 percent are vascular headaches, caused by the excessive dilation of blood vessels in the brain. Migraines are a particularly intense kind of vascular headache. Hangover headaches are another type.

The diameter of cerebral blood vessels is regulated by smooth muscles, which, in turn, are controlled by adenosine. When adenosine levels rise, blood vessels relax and open up. Blocking adenosine receptors with caffeine negates this effect, causing vessels to constrict. This vasoconstriction is relatively minor and usually goes unnoticed, but for people suffering from vascular headaches the reduction of cerebral blood flow is welcomed. This is why caffeine is the active ingredient in many prescription and nonprescription migraine treatments. One of the better-known brands, Cafergot®, contains 100 milligrams of caffeine in each tablet or suppository.

But as a headache remedy, caffeine can be tricky to manage. As we will see in more detail in the next chapter, the body adapts quickly to caffeine. One adjustment is that the muscles controlling the cerebral blood vessels increase their relaxation response to compensate for the increased constriction caused by caffeine. This adaptation goes unnoticed as long as caffeine levels are constantly replenished. But if caffeine intake is suddenly stopped, the increased relaxation is no longer counterbalanced by caffeine. The vessels dilate much more than

ormal, and a throbbing "rebound" headache can ensue. Head-
ches, in fact, are the most common symptom of caffeine with-
rawal. This potential for rebound headaches and the general
ifficulty of accurately self-administering effective doses of caf-
eine are the primary reasons that migraine sufferers are ad-
ised to abstain from caffeine, even though it may bring
emporary relief during an attack.

People suffering from headaches brought on by withdrawal
om caffeine often reach into their medicine cabinet for some
spirin or a non-aspirin pain reliever like Tylenol® or Motrin®.
aking the medication, they may find that their headache van-
hes and they attribute their relief to the aspirin, acetamino-
hen, or ibuprofen in the tablets. But in fact, they may simply
ave given themselves a classic "hair-of-the-dog" cure. Many
rands of aspirin and aspirin substitutes include a significant
ose of caffeine. The usual level is 65 milligrams per tablet,
hich means that the standard two-tablet dose administers a
affeine jolt equal to that in a cup of very strong coffee or two
ups of black tea. This addition of caffeine could be cynically
iterpreted. One might suspect that drug companies add caf-
eine to their pain-reliever products because they recognize
hat headaches are often caused by caffeine withdrawal and
hey understand that the easiest way to relieve drug withdrawal
ymptoms is by the administration of the drug itself.

But there's a more honest—and interesting—explanation for
he presence of caffeine in painkillers. For reasons that remain
bscure, caffeine significantly increases the analgesic effective-
ess of both aspirin and aspirin substitutes such as acetamin-
phen. Data from more than thirty clinical trials involving
nore than 10,000 pain patients unequivocally support this con-
lusion. Most of these patients had pain from postpartum uter-
ne cramping or episiotomies, but some had pain from oral
urgery, headache, or cancer. On average, when the painkillers

were given *without* caffeine, the doses had to be about 40 pe cent larger to obtain the same degree of relief as was obtaine when the analgesics were given *with* caffeine (Laska et a 1984). It is considered safer to give caffeine supplements t patients who are getting large daily doses of pain relievers tha to give them the even higher doses of medicine that would k required without the caffeine.

How caffeine exerts this helping hand is a mystery. One the ory is that the well-being, alertness, and decreased irritabili induced by caffeine somehow counteracts the perception of pain. Some evidence also suggests that caffeine's interferenc with adenosine receptors throughout the body inhibits the pro duction or release of chemicals that cause pain and inflam mation. At the moment, nobody is working to pin down th exact mechanism because so many studies have verified th utility of adding caffeine to pain killers that it is seen as a moc point.

Caffeine and PMS

In the mid-1980s, when she was an assistant professor at Tuft University, Annette MacKay Rossignol was involved in tabu lating the results from a general survey of student health an dietary habits. While poring through the data on health-ris behaviors from exercise to seat-belt use, Rossignol detected a apparent pattern between caffeine and premenstrual syndrome Just for fun, she ran the numbers. The result was a correlatio strong enough to pique Rossignol's interest and launch a re search agenda that lasted for many years.

Premenstrual syndrome is a constellation of symptoms tha can include irritability, anxiety, depression, breast swelling an tenderness, fluid retention, food cravings, and headaches. Ros signol noticed that women with moderate to severe PMS tended to consume the most caffeine, often in the form of soft drinks. Over the next eight years, she and various colleagues conducted a series of ever more refined studies in an attempt to pin down the relationship between caffeine and PMS. She even conducted a study in China to investigate the incidence of PMS among women working in a tea factory who consumed relatively large amounts of caffeine. In this and all of her other studies, she found that the women suffering the most severe PMS also tended to consume the most caffeine-containing beverages, whether tea, soft drinks, or coffee. Rossignol recognized that there might be alternative explanations for the association—for instance, that the effect she was seeing was due not to caffeine but to the extra fluid consumed by the women drinking a lot of caffeine-containing beverages. But a 1990 study controlling for such an effect still found a clear relation between caffeine and PMS.

Work by John W. Phillis of Wayne State University has suggested a possible explanation for Rossignol's findings. Phillis (1989) has found that two of the reproductive hormones that fluctuate monthly in women directly affect adenosine levels in the brain. The two hormones exert opposite effects. Beta-estradiol appears to mimic caffeine's effects: it antagonizes adenosine's natural inhibitory effects and thus produces a kind of mild stimulation that some women characteristically attribute to the late follicular phase of the menstrual cycle—that is, the phase just before ovulation. Conversely, the hormone progesterone *enhances* adenosine's actions. This may explain some of the fatigue and mood depression commonly reported by women in the week after ovulation as progesterone levels increase dramatically. Plummeting progesterone levels just before menstruation may, somewhat paradoxically, account for the feelings of tension, irritability, and anxiety that are sometimes reported during this time. Since caffeine exerts its effects by blocking adenosine, it could—at least theoretically—influence the way these hormone fluctuations affect a woman's physical and mental well-being.

If the theories just outlined are correct—and it is not yet clear that they *are* correct—then women who want to use caffeine to help control the emotional fluctuations of PMS should modulate their caffeine consumption to counteract progesterone's mild depressant effects. About three to four days before they menstruate, women should stop caffeine use altogether and abstain from caffeinated beverages right through their menstruation. After menstruation has finished, they could either remain caffeine-free or drink only small amounts until they ovulate again. Rossignol and Phillis found that many women *did* change their caffeine consumption over the course of a menstrual cycle, but not in the way they expected (Rossignol et al. 1991). Women drank the most caffeine during their menstrual flow, drank less as ovulation approached, and then increased their caffeine consumption—a pattern almost exactly the opposite of that suggested to be beneficial.

Rossignol and Phillis theorize that this pattern of caffeine use may reflect an unsuccessful attempt by the women to self-medicate with caffeine. Women may respond to the unpleasant feelings associated with the peak progesterone levels by increasing their caffeine intake. They maintain this level right through the menstrual flow—at exactly the time, the Rossignol and Phillis studies indicate, they should abstain from caffeine. This may actually make their symptoms worse by leading them to drink even more caffeine in an attempt to overcome their unpleasant symptoms.

As interesting and suggestive as this work is, it should be weighed against one's personal experience. The phenomenon of PMS itself is still not clearly understood, let alone the relationship between caffeine and reproductive hormones. Although the association Rossignol found is well established, it does not prove causation: some other factor may be at work that is unrelated to caffeine. The crucial study, in fact, has yet to be done. This would be a controlled, long-term study of women who receive either caffeine or a placebo in order to see clearly whether caffeine has any impact on premenstrual symptoms. For such a study to be valid, however, the study subjects would have to be ignorant about whether they were drinking caffeine or not. This is hard to do because decaffeinated coffee, tea, and cola generally taste different from their caffeinated versions. Perhaps more important, according to Rossignol, PMS is still not widely considered a phenomenon with public-health consequences, and so raising money for PMS research is extremely difficult.

Tea for Two

During a routine prenatal examination, a fetus being carried by a young woman in the Netherlands was found to have an irregular heartbeat. The woman was admitted to the hospital, where it was learned that she drank more than a quart and a half of caffeinated cola, two cups of coffee, and a mug of cocoa every day. The woman's doctor advised her to abstain completely from these caffeine-containing beverages. Within a week, the fetus's heartbeat had returned to normal. The pregnancy progressed smoothly from then on (Grounds 1990).

This story serves two important purposes. First, it is a reminder that many women drink caffeinated beverages during pregnancy—though usually not to this extreme. A 1983 study of 1,510 women—98 percent of whom regularly consumed coffee or tea prior to becoming pregnant—found that 73 percent continued to drink these beverages during pregnancy (Kurppa et al. 1983). Another study found caffeine in the blood plasma of 75 percent of the newborns tested (Dumas 1982). Second, the story is a reminder that caffeine, like alcohol, freely crosses the placenta. When a pregnant woman drinks a cup of coffee, her unborn child experiences the same degree of stimulation. Women who nurse their infants need to be careful as well because babies lack the liver enzymes needed to break down caffeine. This impact is particularly important for babies born prematurely. In one study, the half-life of caffeine in premature infants ranged from 41 to 231 hours. The average adult half-life, in contrast, is 5 to 6 hours.

The big question, of course, is whether caffeine really harms fetuses or nursing infants. Animal studies using levels of caffeine far higher than any consumed by humans clearly demonstrate that caffeine can be a teratogen—that is, an agent capable of causing birth defects. It was partly in response to such studies that in 1980 the FDA advised pregnant women to reduce their intake of caffeine to a minimum.

Since that time, other animal studies using more typical caffeine doses have shown an association between caffeine intake and lower birth weights, as well as increased incidences of still births and miscarriages. A number of theories, none proved, have been put forward to explain these effects (Eskenazi 1993). For instance, caffeine doses as low as 200 milligrams (the amount in two cups of coffee) decreases placental blood flow. Caffeine also increases the force of contraction in the fetal heart and decreases the levels of the most abundant kind of estrogen: a reproductive hormone called estradiol. Epidemiological studies on humans have supported the view that consumption of moderate to high levels of caffeine during pregnancy (more than 300 milligrams a day) slightly increases the risk of spontaneous abortion, intrauterine growth retardation (low birth weight), and microcephaly—a condition in which the head and brain are abnormally small.

The evidence of developmental problems arising from lower levels of caffeine consumption is less clear. Two studies, both rigorously conducted and published in a leading medical journal, yielded very different findings on the impact of low levels of caffeine consumption. Although both found evidence for intrauterine growth retardation among women who consumed more than 300 milligrams of caffeine a day, one study (Mills et al. 1993) found *no* harmful effects for doses lower than 300 milligrams, while the other (Infante-Rivard et al. 1993) found a significantly increased risk for spontaneous abortion among women who consumed doses as low as 163 milligrams a day during the first trimester.

Only a few studies have been conducted on children of mothers who drank caffeinated beverages during pregnancy. One showed no effects of caffeine on infant and child neuro-development (Barr and Streissguth 1991); another found maternal caffeine consumption to be responsible for poor neuromuscular development and greater arousal and irritability among babies (Jacobson et al. 1984). In light of these and many other studies offering conflicting or difficult-to-interpret data, many doctors err on the side of caution and advise women who are either pregnant or planning to conceive to abstain from caffeine (Eskenazi 1993). The same advice is given to women who breast-feed, since caffeine readily passes into breast milk, exposing an infant during a period of rapid neurodevelopment.

To the Bone

Some recent studies have given rise to popular concern about caffeine's effects on bone density and osteoporosis, a thinning of the bones that tends to occur later in life, especially among women. One such study resulted in headlines that, ironically, may have exacerbated these fears even though the study results themselves did not directly implicate caffeine as causing bone loss.

The researchers studied 980 women and found a small decrease in bone density among women over the age of fifty who had drunk at least two cups of coffee a day for many years (Barrett-Connor et al. 1994). But the point that got lost in most media reports was that the loss was found only among women who *hadn't* drunk at least 8 ounces of milk a day on a regular basis. Among those women who had consumed at least a glass of milk a day, there was no increased risk of bone loss even though they drank coffee. In other words, these results suggest that it's not coffee drinking per se that seems to be responsible for the loss of bone density, but the reduction in calcium intake because some women drink coffee *instead of* milk. The moral, then, is that *everyone*, whether they drink coffee or not, should consume adequate amounts of calcium.

Other studies have shown that neither the absorption of calcium from foods nor the excretion of calcium from the body is affected by caffeine (NIH 1994), leading most researchers to conclude that caffeine has no significant impact on general bone density or the disease of osteoporosis.

Decaf Jitters

The glory years of decaffeinated coffee appear to be over. Back in 1985, about 17 percent of the adult coffee-drinking population drank decaf (National Coffee Association 1993). That was up considerably from the 4 percent figure in 1962. But since 1985, fewer and fewer people take their coffee "unleaded." By 1993 the percentage had dropped to 12.1 percent, and the trend shows no signs of halting.

The move away from decaf may have been spurred by results

om two separate studies hinting that decaf—but not regular
offee—might increase the risk of heart disease. The first
udy, conducted in 1989 at Stanford University, showed that
a group of 181 men, those drinking decaf experienced a 7
ercent rise in their levels of low-density lipoproteins, the so-
alled bad cholesterol. Although statistically significant, there
ere reasons to doubt that the LDL rise was related to the
ecaf. For instance, the data didn't show any relationship be-
ween the amount of decaf consumed and the degree of the
holesterol rise. Finding such a dose response would have made
he case stronger. Even the director of the laboratory in which
he study was done said that he had no plans to change his
abit of drinking three cups of decaf a day (Lehman 1989).

A second study tracking the health of 45,589 doctors for two
ears found *no* association between caffeine and heart dis-
ase—good news for regular-coffee drinkers (Grobbee et al.
990). But the data turned up a "marginally significant" in-
rease in heart disease among the men who drank four or more
ups of decaf a day. The study authors interpreted their find-
ngs cautiously. They noted that since the number of decaf
rinkers was relatively small, the results were subject to larger
nargins of error than the data for regular coffee. They also
ointed out that they had studied only men, and thus their
indings might not apply to women. Finally, as with the Stan-
ord study, the effects found might have been attributable to
ome unexamined difference among the men in the two
roups.

Another popular concern about decaf is related to the de-
affeination process itself. Some fear that substances used in
he process remain in the beans and could pose a health threat.
uch fears are probably unwarranted.

There are two basic ways to remove caffeine from coffee

beans. In the "water process," green coffee beans are soake
in hot water for ten minutes to two hours. In addition to leech
ing out most of the caffeine, this process removes most of th
compounds that give coffee its flavor and body. Make coffe
from the beans at this stage, and a truly wretched brew wi
result. That's why manufacturers take great pains to return a
much of those lost flavor compounds as they can. After th
soaking, the caffeine-laced water is drawn off and the caffein
is removed, either with a caffeine-specific solvent (such a
methylene chloride or ethyl acetate) or by passing the wate
over acid-treated carbon filters to which the caffeine binds. Th
liquid is then returned to the beans, which reabsorb some o
the flavor compounds. After this step, the beans are dried an
shipped to roasters.

So-called solvent processing is more direct. Here, the gree
coffee beans are washed with a caffeine solvent (again usuall
methylene chloride but sometimes ethyl acetate) in tubs o
rotating drums. The caffeine is then filtered from the liquic
solvent. Because this process is relatively fast and because sol
vents are more specific than water in their action, more of th
delicate flavor compounds are usually retained in solvent
processed beans.

After the caffeine-laden solvent is removed, the beans ar
processed with steam to remove residual solvent. This remova
is very effective because methylene chloride and other caffein
solvents evaporate at temperatures between 100 and 120 de
grees. In comparison, steam is 212 degrees and coffee bean
are roasted at temperatures of 350 to 425 degrees. Virtually n
solvent remains by the time the coffee reaches your cup, which
is why the Food and Drug Administration in 1985 ruled tha
there is no risk in drinking solvent-processed decaffeinated
coffee.

The Last Drop

The effects of caffeine examined in this chapter are all *acute* effects; that is, they take place shortly after ingestion. Whether the effect is heightened athletic performance or an increased urge to go to the bathroom, these changes are all relatively short-lived. Once the caffeine has been metabolized, the effect disappears.

But people tend to drink caffeine on a regular basis over long periods of time—often the greater part of a lifetime. What effects might such long-term consumption have on the body, and how does the brain, in particular, respond to this situation? In the section on headaches, we saw that blood vessels in the brain quickly adapt to caffeine and that this can lead to rebound headaches if caffeine intake is stopped abruptly. In the next chapter, we'll take a closer look at this phenomenon as it applies to the brain as a whole, and we'll look at the question of caffeine's addictive potential.

It was purely the feeling that had captivated me, made me sacrifice everything to it, gladly, joyfully. It was a seashell's pristine whisper in my ear, warm sun rising in my heart, fireflies winking in the nerves.

—Will Bohnaker, Haunts of the Aardwolf,
on the allure of caffeine

Hooked

Caffeine Anonymous

"We were all standing there twitching," Mike said, recounting a time standing in line waiting for coffee at a local café. "Everyone was saying 'Come on, let's go, let's go. What's the holdup?' We were like heroin junkies" (Richards 1995).

Mike is a member of the nation's first chapter of Caffeine Anonymous, a support group based in Portland, Oregon, that is modeled on Alcoholics Anonymous. Founded in April 1994, the group remains tiny. Meetings typically consist of between ive and eight people seated in a spare room in St. Stephen's Episcopal Church. Size notwithstanding, members have strong feelings about the substance that bedevils them: step 1 of the group's twelve-step program calls on members to "admit that we are powerless over caffeine and that our lives have become unmanageable." For the group's cofounder, Marsha Naegeli-Moody, the word "addiction" exactly describes the compulsion she felt at the height of her caffeine consumption. She was out of control, she said, knocking back up to ten cups of coffee a day. She predicts that others in the grip of caffeine will form similar groups in the future to help those who want to kick their habit. "In five years coffee is going to be treated just like nicotine," she said in a newspaper interview.

Others find this attitude extreme. "Addicting? Hogwash," grumbled an on-line participant in a 1995 Internet discussion on the Portland group. "Habit-forming, maybe. But let's not lump caffeine in with heroin, crack, and alcohol." At first glance, this kind of skepticism toward claims of caffeine addiction, or "caffeinism," seems reasonable. In many ways, caffeine is in a different league from other recreational drugs. For one thing, caffeine's power to intoxicate is relatively weak. The buzz from a couple of cups of coffee is mild compared with a typical hit of cocaine or amphetamine and is trivial compared with a dose of LSD.

The effect of caffeine is often so subtle that it is impossible to tell if someone has consumed it or not—a fact corroborated by accident in one of the classic experiments on human reactions to alcohol and caffeine. In this study, subjects were given alcohol, caffeine, or a placebo beverage. An examiner, who was not told what the subjects had consumed, then tested the volunteers on everything from their reaction time to their emotional state (Nash 1962). It turned out that the examiner could easily tell when a volunteer had consumed alcohol, but the examiner could *not* tell when subjects had consumed caffeine, even though the dose was significant: 300 milligrams.

The subtlety of caffeine's effects is also evident in the fact that it is typically used to "normalize" rather than "intoxicate." Caffeine is helpful in sustaining mental functioning under conditions of fatigue or boredom, such as that experienced by late-shift workers, students cramming for an exam, or long-distance truck drivers. The only other recreational drug used in this way is nicotine, which is also seldom used for outright intoxication. Few would hesitate to fly in a plane being piloted by a coffee drinker or cigarette smoker, whereas an alcohol-guzzling or cocaine-snorting pilot would be worrisome indeed.

Caffeine, of course, differs from nicotine in that it is less hazardous to one's health. In fact, caffeine is arguably the safest recreational drug. That, in any case, is the bottom line of a great many studies into the matter. The effects of caffeine on such things as breast cancer, bone loss, pancreatic cancer, colon cancer, heart disease, liver disease, kidney disease, and mental dysfunction have been examined in exhaustive detail and, to date, *no* clear evidence has been found linking moderate consumption of caffeine (the equivalent of three to four cups of coffee daily) with these or any other health disorder (Chou 1992; Goldstein 1994; Gordis 1990; Grobbee et al. 1990). Still, caffeine is not aspirin. Certain individuals may be unusually sensitive to caffeine even with low doses and may experience adverse effects such as increased anxiety, or cardiac abnormalities such as palpitations or heart arrhythmias. And, as previously noted, caffeine freely passes the placental barrier and also easily enters breast milk, which means that abstinence is probably the safest choice for pregnant or breast-feeding women.

Balanced against these specific warnings is the fact that far from killing people, caffeine undoubtedly *saves* lives—though this is difficult to prove. Statistics cannot be tabulated to determine how many drivers might otherwise have fallen asleep at the wheel had they *not* downed some coffee or a tablet or two of No-Doz® before setting out. This, of course, stands in stark contrast to the estimated 400,000 Americans who die each year from tobacco use, and the 100,000 whose deaths are attributable to alcohol (Glass 1994).

And yet, despite its safety and mildness relative to other recreational drugs, caffeine still unquestionably alters brain function. These alterations trigger adaptive changes in the brains of even casual users, resulting in such hallmarks of drug addiction as tolerance, dependence, craving, drug-seeking behavior, and, after cessation, withdrawal symptoms. As with alcohol, a minority of users find caffeine to be exceptionally attractive; they crave it strongly, ingest ever larger amounts, and suffer worse withdrawal symptoms than do most users. This is why the members of Caffeine Anonymous cannot be dismissed as overzealous handwringers, nor can their claim of caffeine's addictive potential be swept aside as "hogwash."

"Caffeinism" is certainly a much less pressing societal problem than alcoholism or nicotine addiction. It is highly unlikely that membership in Caffeine Anonymous will ever rival that in AA. But caffeine's ability to induce drug reactions that resemble those experienced by addicts of truly potent drugs is widely unappreciated.

Setpoint

The variation in people's physical responses to caffeine is impressive. Some people can drink several cups of coffee after dinner, fall soundly asleep an hour later, and sleep peacefully until morning. Others find that even one cup of coffee early in the day induces a fitful night's sleep. Likewise, the caffeine in a single cup of tea makes susceptible individuals anxious and unpleasantly nervous, while for others caffeine is both a relaxant and a mood-enhancer.

Some of this variation is undoubtedly rooted in genes. As we've seen, the genes that give us unique faces and fingerprints also give us unique brains. No drug that acts on the brain, therefore, is going to act exactly the same way in everyone. No one knows what kinds of individual differences lie at the bottom of people's divergent sensitivities to caffeine. But there are some hints. For example, people might differ in the number and distribution of their adenosine receptors. These receptors are manufactured according to blueprints stored in DNA. This information must be translated, the manufacturing processes carried out, and the finished receptors shepherded to their proper locations in nerve-cell membranes. All these steps require exquisitely delicate molecular controls, and variations in any of the steps could result in a person ending up with more or fewer adenosine receptors than normal. A person with an above-average endowment of adenosine receptors—that is, someone with more targets for caffeine to hit—might be hypersensitive to caffeine. Conversely, people with fewer-than-normal adenosine receptors might be unusually *in*sensitive to caffeine.

Again, this line of reasoning is purely speculative. No one yet knows how much people vary in the quantity and quality of their adenosine receptors, nor is it completely clear what effects such variations have in terms of behavior. In reality, genes probably affect caffeine sensitivity in dozens of ways, most of them not yet even guessed at. But genetic variation isn't the only—and maybe not even the most important—reason people differ so much in their reactions to caffeine (Dews 1984). Although it may sound like circular reasoning, one of

ie reasons people differ in their response to caffeine is that
eople differ in their *consumption* of caffeine. Consumption,
i other words, can radically affect sensitivity.

Many habitual drinkers of caffeine-containing beverages find
hat they must increase their dose to achieve the preferred
egree of stimulation. Lying behind this phenomenon is the
rain's remarkable plasticity. To an extent far greater than any
ther organ, the brain adapts to changing conditions. It has to.
n addition to being the seat of consciousness and awareness,
he brain controls heartbeat, breathing, and other life-support
ystems. Wild fluctuations in brain activity owing to changing
nvironmental conditions would thus put the rest of the body
t severe risk. Shaped by millions of years of such selective
ressure, the human brain today comes equipped with dozens
f mechanisms designed to tightly regulate the level of brain
ctivity. Like thermostats, they constantly adjust such things
s neurotransmitter release and receptor sensitivity to compen-
ate for perturbations from the environment.

The situation is analogous to another regulator in the
ody—the one controlling weight. As most dieters know from
ard experience, the body has a "setpoint"—a weight that it
trives to maintain despite fluctuations in food intake. This
etpoint varies between individuals and is fundamentally gov-
rned by genes (Leibel et al. 1995). Research has shown that
vhen calories are cut, metabolism slows in compensation. If
xcess calories are consumed, metabolism speeds up in an effort
o burn off the extra calories and bring body weight back to
he setpoint.

The same principle applies to the brain's setpoint. Attempts
o rev up brain activity (for instance, with caffeine) are met
vith a countervailing response that reduces brain activity. At-
empts to *lower* brain activity (as with alcohol) are met with
he opposite response. The brain constantly strives to maintain

its genetically governed setpoint of activity, even if this means going to extraordinary ends to achieve it. This flexible response to any drug, whether recreational or therapeutic, is called tolerance. Longtime heroin users, for instance, have been observed to require *ten thousand times* the dose they injected when they began their habit. Their brains adapt to heroin to such an extent that they inject themselves with quantities of this narcotic that could kill a person not tolerant to heroin. In contrast, the tolerance achievable by even the heaviest coffee drinker rarely requires more than ten to fifteen times the caffeine a first-time drinker might consume (Goldstein 1994). That hardly means, however, that tolerance is a trivial issue.

Up Escalator

Research with drugs such as heroin has revealed that one way the brain responds to drug-induced perturbations is to change the number of receptors in the affected neurotransmitter system. This now appears to be one way the brain reacts to caffeine.

As we've seen, caffeine plugs adenosine receptors, thus blocking their normal ability to slow the brain down. This blockage is detected via an unknown mechanism and triggers the creation of *more* adenosine receptors. It's as though the receptors were antennae picking up a steady radio signal; when the signal suddenly weakens, more antennae are added to the system to compensate. This adaptive process of increasing receptors is called up-regulation, and it is a common brain response to any drug that blocks a specific circuit or neurotransmitter. The opposite response, called down-regulation is typically seen in reaction to drugs—such as heroin—that directly *stimulate* neurotransmitter receptors.

A number of studies have shown that caffeine and other

methylxanthines up-regulate adenosine receptors in many tissues, including the brain (Fastbom and Fredholm 1990; Sanders and Murray 1988). Up-regulation thus may be responsible at least in part for tolerance to caffeine, though, again, how this happens is not clear, and studies on whether caffeine causes adenosine-receptor up-regulation have yielded contradictory results (Kaplan et al. 1993; Zielke and Zielke 1987).

It appears that chronic caffeine use may cause up-regulation or down-regulation of other neurotransmitter systems as well. Recent experiments with mice revealed the expected 20 percent up-regulation of A_1 adenosine receptors, but the scientists also found surprising changes in receptor densities for many other important neurotransmitters (Shi et al. 1993). Receptors for norepinephrine (a hormone similar to adrenaline) were reduced. Densities of certain serotonin receptors were increased, as were densities of acetylcholine receptors. And a striking 65 percent up-regulation of GABA receptors was observed. These results suggest that caffeine indirectly affects many neurotransmitter systems through its direct effects on adenosine receptors. The added firing in many brain circuits owing to caffeine intake undoubtely increases or decreases the release of dopamine, serotonin, and other important neurotransmitters. It is too early to say what effect these secondary alterations have on behavior, though they may turn out to be important components of the general experience of being "wired" on caffeine.

Tolerance to caffeine sets in relatively quickly. Animal experiments with large doses of caffeine have induced tolerance in as little as three days (Daly 1993). Tolerance in humans develops a bit more slowly, probably because humans do not ingest the large amount of caffeine typically administered to test animals. Still, humans generally become tolerant to a given dose of caffeine—whether a single can of soda or ten cups of coffee—in a week to twelve days (Regestein 1995).

This tolerance can be remarkably complete; that is, the brain's ability to compensate for caffeine can be so effective that tolerant users experience very little, if any, true stimulation by their customary dose. This was demonstrated by a particularly rigorous experiment in which thirty-two healthy volunteers participated in a monthlong study of the subjective effects of caffeine (Evans and Griffiths 1992). Half the volunteers received caffeine (in capsule form), and half received a placebo. The people consuming the caffeine demonstrated tolerance in several ways. On so-called forced-exposure days in which *all* participants received caffeine, those who had been taking the placebo showed much greater effects from the caffeine than those who had been taking caffeine regularly. The scores for the placebo group on such things as tension, anxiety, jitteriness, and the perceived strength of the drug were often several times greater than the scores of those in the caffeine-tolerant group.

Most telling were observations made during the portion of the experiment in which ingestion of either a placebo or caffeine was held constant for eighteen days. The two groups were given a battery of tests to rate everything from their mood to their physical health. Remarkably, even though the people in the caffeine group were consuming 900 milligrams a day, the average scores for both groups were virtually identical. The participants reported *no* significant differences in such things as energy, alertness, irritability, talkativeness, tension, and depression/dejection. Evidently, the brains of the caffeine consumers had adapted fully and relatively quickly to caffeine—to the extent that they were "normal," at least compared with those of their non-caffeine-consuming peers.

This raises an obvious question: If people become tolerant to caffeine in a matter of days and thereafter derive essentially no stimulation from the drug, what accounts for the enormous popularity of caffeine-containing beverages (other than the fact that many people find them delicious) and for the distinct sense by users that they are, in fact, being stimulated? One possibility is that some parts of the brain may *not* become tolerant to caffeine. Even heavy long-term users may thus be feeling some kind of "buzz" through the general muffler of tolerance (Nehlig et al. 1992). Evidence supporting this idea is, however, still quite tenuous, and even if it's true, the effect is likely to be rather subtle. A more likely explanation for why people continue to consume caffeinated beverages long after tolerance has been established can be found by looking at the flip side of the phenomenon of tolerance: withdrawal.

On the Rebound

Since caffeine is relatively inexpensive and widely available, the dose escalation induced by tolerance is seldom burdensome. Nobody must resort to crime to support their habit, nor do they need to rely on back-alley dealers to supply their daily fixes. Of course, tolerance to *very* heavy doses of caffeine may be problematic for health reasons. Both coffee and tea, for instance, are fairly acidic beverages, and some people find that ingestion of large amounts irritates their stomach. But by and large, tolerance isn't even noticed as long as circulating levels of caffeine are kept stable. The problems come when those levels drop, at which time the brain's delicately balanced see-saw of neurotransmitters and receptors tips radically. Without the "weight" of caffeine to push against, the brain goes overboard. The result is withdrawal: a constellation of physical and psychological symptoms that in the case of caffeine can range from imperceptible to intensely unpleasant—though caffeine withdrawal is never lethal the way withdrawal from alcohol or heroin can be.

By far the most common symptom of caffeine withdrawal i̇
headache—a fact that has only recently been accepted by the
medical community. This acceptance grew out of the long
observed phenomenon that many patients given general anes
thesia experience a headache when they come to. Such
postoperative headaches have traditionally been considered ȧ
unavoidable side effect of the anesthesia itself. But in 1989
three doctors at Hammersmith Hospital near London ques
tioned this assumption (Galletly et al. 1989). They decided ṫ
test another theory about the origin of the postoperative head
ache: that it is due to caffeine withdrawal initiated by the stan
dard requirement that patients undergoing elective surger
involving general anesthesia abstain from both food and caf
feinated beverages prior to their operation.

The doctors investigated their hunch by having 142 ran
domly selected patients fill out a questionnaire after they re
covered from their anesthesia. The survey asked how the
patients felt as well as what their typical intake of caffeine wa
before the surgery. The results showed that the doctors were
on to something: the more caffeine patients consumed, the
more likely they were to experience headache after anesthesia
Seventy-three percent of the patients consuming more than
100 milligrams of caffeine a day experienced headache; at the
other end of the spectrum, *none* of the six patients who con
sumed *no* caffeine prior to surgery had a headache after anes
thesia.

That it has taken so long for this seemingly straightforward
association between caffeine intake and postoperative head
ache to be accepted says volumes for the popular perception
of caffeine. Until recently, caffeine was not considered a drug
by either the general public or doctors, hence the idea that
sudden cessation of caffeine could precipitate withdrawal
symptoms wasn't considered.

Since that 1989 study, several other major investigations have unequivocally demonstrated the reality of caffeine withdrawal symptoms (Silverman et al. 1992, Strain et al. 1994). Here are the typical symptoms of caffeine withdrawal in rough order of their occurrence in the general population:

- Headache
- Depression
- Fatigue
- Lethargy
- Irritableness
- Increased muscle tension
- Nausea
- Vomiting

Silverman's study included some direct quotes from people experiencing caffeine withdrawal:

"I felt like I had the flu, a severe headache, extreme fatigue."

"I felt sad, uncertain about the future, a general feeling of glum."

"I couldn't concentrate even when I had to do those tests. I'm basically not a low person: I was mildly sad and depressed."

The most extreme response came from a woman who said, "I had a severe headache that progressed into vomiting, flu-like symptoms. I can only compare that sickness to the radiation and treatment (radiation and chemotherapy treatment for cervical cancer) of the past year. It was as bad as that."

Not surprisingly, the studies of caffeine tolerance and withdrawal have found wide variations in subject responses. Some people, even heavy consumers, report no withdrawal symptoms at all, while others suffer severe headaches and other unpleasant symptoms even though they consumed as little as one cup of coffee a day prior to abstention. In general, withdrawal symptoms begin within twelve to twenty-four hours after the last use and peak anywhere from twenty to forty-eight hours after caffeine consumption stops (Hughes 1992). Withdrawal symptoms then typically taper off, but it usually takes a full week for a return to normal.

These timings are significant. Most regular consumers of caffeine are in the first stages of caffeine withdrawal when they wake up in the morning (assuming they didn't drink coffee just before going to bed the previous night). For this reason, caffeine users, in general, are likely to feel more tired, irritable and groggy in the morning than people who abstain from caffeine. Also, if morning caffeine intake is skipped, a headache is likely later in the morning or early that afternoon. Drinking caffeine, of course, quickly alleviates these withdrawal symptoms, just as the classic "hair-of-the-dog" nip of alcohol "cures" a hangover. It's not surprising, therefore, that the morning ritual for many caffeine consumers involves getting caffeine into the bloodstream as quickly as possible. The presence of morning withdrawal symptoms also explains why the first cup of coffee, tea, or soda can give the most pronounced boost to mood and energy: the effect is more obvious because of the contrast with the feelings of lethargy and depression associated with withdrawal.

Thanks in part to the studies just cited, the reality of caffeine withdrawal is coming to be appreciated by both physicians and the lay public. Physicians, for instance, have been advised to ask about caffeine use when patients complain of symptoms such as headaches, depression, fatigue, and drowsiness (Hughes 1992). Another consequence of the growing appreciation for caffeine withdrawal is the recognition that, for a minority of users, caffeine may have addictive potential.

ependence

olerance and withdrawal are the classic signs of physical de-
ndence on a drug. But physical dependence isn't the same
ing as addiction. Morphine and Valium, for example, regu-
rly produce physical dependence. The brain down-regulates
e opiate and GABA receptors that are the immediate targets
these two drugs, resulting in a need for larger and larger
oses as time goes on. But opiates and the benzodiazepines
ave important medical uses in the management of pain and
axiety. Their controlled use by patients who have been care-
lly screened, well informed, and closely monitored by physi-
ans does not constitute addiction even though such patients
early become physically dependent on the drug after a period
regular use.

So what *does* constitute addiction? The arbiter of such things
ese days is the latest version of the *Diagnostic and Statistical
Manual of Mental Disorders* (DSM-IV) of the American Psy-
niatric Association. The DSM-IV defines two kinds of problem
lationships with drugs: substance abuse and the more serious
ubstance dependence. This latter category is what most people
ould call addiction, though the DSM-IV avoids that word
ecause it is so heavily freighted with moral and emotional
onnotations.

Substance dependence is characterized by using a substance
larger amounts or for a longer period of time than intended;
peated unsuccessful efforts to cut down or control use; use
a substance to relieve or avoid withdrawal symptoms; or a
attern of compulsive drug-taking that persists despite clear
ocial, psychological, physical, or occupational problems related
the drug. For decades, no one thought that caffeine was a
otent enough drug to cause the serious problems associated

with substance dependence. But that view has been revised light of a recent study by Eric Strain and his colleagues at th Johns Hopkins University School of Medicine. The researche wanted to see whether they could find caffeine users wh would qualify as substance dependent under the DSM-IV de inition (Strain et al. 1994). Newspaper ads were used to loca people who thought they were psychologically or physically d pendent on caffeine. Out of ninety-nine people screened f the study, sixteen were diagnosed as caffeine dependent aft undergoing a battery of evaluations.

The average daily consumption of caffeine of these partic pants was 357 milligrams, which is somewhat but not striking higher than the 280-milligram average consumed in the Unite States. The actual daily amount consumed by the individua ranged from a low of 129 milligrams to a high of 2,548 mill grams. Half the subjects got their caffeine from coffee, 44 pe cent drank caffeine-containing soda, and one participant dra tea. Fully 81 percent of the caffeine-dependent subjects sa they had made unsuccessful efforts to cut down or control the use, and 94 percent said they consumed caffeine despite a pe sistent or recurrent physical or psychological problem relate to its use. Almost half of the subjects reported some physic condition such as heart palpitations or gastrointestinal pro lems that had led their physicians to recommend reducing eliminating caffeine consumption—and in each case the su jects were unable to do so.

In a second phase of this study, the researchers tested eleve of the sixteen caffeine-dependent subjects for withdraw symptoms. Interestingly, two of the eleven showed no wit drawal symptoms at all, though they told the researche that they had experienced such symptoms when they tried quit in the past. This finding is further evidence of the wi variability in withdrawal effects from person to person and even among individuals themselves from one episode to the next.

A variety of impairments were reported by subjects when they were in withdrawal, including

- Missing work owing to bouts of vomiting
- Making multiple costly mistakes at work
- Going home from work early to sleep
- Inability to complete schoolwork
- Screaming at the children
- Cancellation of a son's birthday party
- Being too tired to do household chores

In addition to underscoring the seriousness of caffeine withdrawal and settling the question of whether some people really do qualify as "caffeine addicts," the Johns Hopkins researchers uncovered an intriguing pattern in the data. Fifty-seven percent of those diagnosed as caffeine dependent had earlier been diagnosed as suffering from either alcohol abuse or alcohol dependence. In addition, seven of the sixteen subjects had a past diagnosis of manic-depressive disorder or major depression. The researchers suggested that this clustering of caffeine dependence with alcohol abuse and/or mood disturbances deserves further study. Do some people turn to caffeine and other drugs because they are trying to self-medicate an underlying mental dysfunction? Does caffeine use by such people somehow exacerbate or initiate other substance-abuse problems? Is there an "addictive personality" predisposed to dependence on many types of drugs? These are just some of many questions awaiting a fuller understanding of the biological and psychological underpinnings of addiction in general.

Caffeine Paradox

Consider the following real-life case studies:

- A thirty-five-year-old office worker who sleeps twelve hours a night, falls asleep every time she watches television, and stays in bed all day on Sunday, even though she drinks ten cups of coffee and two liters of cola a day

- A fifty-two-year-old secretary who oversleeps regularly and feels unbearably groggy in midafternoon despite her daily consumption of six or seven cups of coffee as well as a prescribed stimulant

- A forty-five-year-old cabinet maker who wakes up groggy, takes a nap every day, falls asleep over meals, and has outbursts of temper, even though he drinks six or seven cups of coffee and day and supplements them with caffeine pills.

In each of these cases the disagreeable symptoms disappeared when the patients stopped taking caffeine or other stimulants (Caffeine 1990). For these people, in other words, caffeine was acting as a depressant, not a stimulant: instead of energy, motivation, and heightened mood, they experienced lethargy, sleepiness, and depression.

There are several possible explanations for this paradoxical effect of caffeine, according to Quentin Regestein, the director of the Sleep Clinic of Brigham and Women's Hospital in Boston, where these patients were treated. For instance, the heavy caffeine use by these people might have interfered with their sleep at night to such an extent that they were simply exhausted during the day—so much so that more caffeine could not overcome their torpor. Or perhaps these people were sim-

ly hypersensitive to the depressant effects of very high doses
of caffeine. Regestein believes that the answer lies buried in
the complicated variations of receptor profiles and neurotransmitters among individuals. "This is why medicine isn't a sci-
nce," he says. "We just don't know what's going on with these
people. All we know is that they improve when caffeine use is
topped."

Another example of a paradoxical effect, similar to the one
observed by Regestein, involves the use of stimulants in the
treatment of attention deficit hyperactivity disorder (ACHD).
The serendipitous discovery in the 1930s of the calming influence of certain stimulants on the behavior of some children
diagnosed as hyperactive led to a search for alternative stimu-
ants that would be effective, non–habit forming, and inexpen-
ive (Gittelman 1983). Caffeine was one of the first such drugs
o be studied.

Of the seven controlled studies of caffeine and hyperactivity,
ve failed to detect any advantage of caffeine over a placebo;
wo reported significant improvement. The positive results of
hese latter two studies have been questioned because of their
elatively small sample size. Since no study reported that caf-
eine was *worse* than a placebo, a consensus emerged that
affeine probably has a weak but clinically unsatisfactory ther-
peutic impact on children labeled as having ADHD.

The neurological mechanisms behind the effects of stimu-
ants such as caffeine or Ritalin® (a popular drug treatment for
ADHD) are not yet known. From a theoretical point of view,
owever, this kind of paradoxical effect isn't inherently mys-
erious. Many of the neurons in the cerebral cortex—the seat
of "higher" functions such as rational thought, speech, and
reativity—*inhibit* the functioning of other parts of the brain.
When such neurons fire, they *dampen* activity elsewhere, which
s, apparently, a critical function in a healthy brain. Increasing

the firing rate of these cortical neurons by stimulants, in other words, could have the paradoxical effect of turning down the "volume" in other brain circuits, perhaps allowing for the increased attention span and ability to concentrate that is sometimes observed when hyperactive children are given stimulants.

Since most people do *not* experience a depression of cortical function when they take stimulants such as caffeine, the kind of paradoxical inhibition seen in some sleep-disorder patients and hyperactive children may be due to an underlying neurological difference in their brain chemistry. Much current research is aimed at testing this idea in the hope of finding more effective treatments for both problems.

Variation Redux

In this chapter, we've seen that the brain quickly adapts to caffeine—as it does to other drugs—because it continually strives to maintain a "setpoint" of neurological activity. Such adaptations lie behind the phenomenon of tolerance, and tolerance explains why regular users of caffeine experience a reduced "kick" from their standard dose within a matter of days. Tolerance also explains why many people experience withdrawal symptoms when they stop drinking caffeineated beverages. Having adapted to caffeine, the brain "rebounds" in its absence, producing a constellation of unpleasant symptoms such as headache, fatigue, depression, and irritability.

Tolerance and withdrawal are signs of physical dependence, which is *not* the same as either substance abuse or substance dependence (addiction). Although most caffeine users are physically dependent to one degree or another, many additional factors must be weighed before the labels of abuse or addiction can be used. Only for a minority of users can caffeine be termed "addictive."

We end our exploration of caffeine, therefore, exactly where we ended our look at alcohol: contemplating the range of human biological diversity. The practical consequence of this diversity is the nonexistence of blanket rules or guidelines for caffeine use. In the end, the best answers come from personal experimentation with varying doses of caffeine to see how this drug interacts with one's unique biochemistry. As with alcohol, the information presented here about how caffeine works is probably most helpful as a baseline against which to gauge one's personal experiences.

If it weren't for the coffee, I'd have no identifiable personality whatsoever.

—David Letterman

10 Better Living Through Chemistry

The Missing Link

Up to this point we've ignored a rather important fact about alcohol and caffeine: that many people end up with *both* drugs circulating through their brain at the same time. This sometimes results from the consumption of a dual-drug beverage such as Irish coffee (whiskey and coffee) or rum and Coke. More commonly, beverages containing alcohol or caffeine are consumed in close temporal proximity to one another, as when a meal begins with wine and ends with espresso, or when a martini follows a long day of slugging coffee at the office.

What happens under these circumstances? How do alcohol

and caffeine interact? Even though they are opposites in many ways, they clearly don't simply annihilate each other on contact like matter and antimatter. People who consume four Irish coffees in rapid succession are anything but sober. But they will not be feeling either purely intoxicated or purely wired. What manner of inebriation will they be experiencing, and what neurobiology supports it?

For a long time, it was believed that caffeine and alcohol went their separate ways in the brain. It was thought that they worked on fundamentally different brain circuits and neurotransmitter systems and that they did not, therefore, directly antagonize each other's actions—an idea enshrined in the standard advice that if you try to sober up a drunk with caffeine you'll simply end up with a wide-awake drunk. This assumption is sound, even though the premise on which it's based has been proved wrong.

Research has shown that there *is* a direct link between the actions of alcohol and caffeine. The two drugs counteract each other's influence on one of the brain's important neurotransmitter systems, which means that, to a limited extent anyway, caffeine and alcohol *can* neutralize each other. The first hints of this relationship arose from alcohol studies using the mice we met in Chapter 5: long-sleep mice, which become comatose on low doses of alcohol, and short-sleep mice, which tolerate relatively high doses and nap only briefly when finally overcome.

In the early 1980s, neuroscientists were trying to pin down the neurological basis for these markedly different reactions to alcohol. It proved to be quite difficult. Neurotransmitter system after neurotransmitter system was examined and found to be essentially identical between the two strains of mice. Finally, in 1984, a significant neurotransmitter difference *was* found in the then relatively obscure adenosine system.

William Proctor and Thomas Dunwiddie in the Department

of Pharmacology at the University of Colorado's Health Sciences Center discovered that short- and long-sleep mice responded very differently to drugs affecting adenosine. For instance, a drug called L-PIA, which mimics adenosine, caused the long-sleep mice to become *very* sleepy and lethargic, while having little effect on the short-sleep mice (Proctor and Dunwiddie 1984). When the researchers gave the mice theophylline, which antagonizes adenosine receptors, the long-sleep mice were 61 percent more active than usual. The short-sleep mice, in contrast, showed no increase in activity after the injection. In short, these two strains of mice, which react very differently to alcohol, also reacted very differently to drugs affecting adenosine. These were striking results because they implied a strong neurochemical connection between the two most popular drugs on the planet. Despite its implications for everyday consumers of alcohol and caffeine, however, the study findings didn't make headlines. This fundamental relationship between the actions of alcohol and caffeine has thus remained virtually unknown to all but a few neuroscientists who specialize in adenosine.

Among those scientists, however, the paper set off a search for the molecular mechanisms underlying the observations in mice. The obvious place to start was to see whether alcohol directly affects adenosine levels in the brain. Initial reports have been positive: when neurons are exposed to alcohol, adenosine levels increase in their vicinity. Since adenosine often depresses neuronal firing, its liberation by alcohol would contribute to the sedation and lethargy experienced by people who drink moderate to heavy doses.

Exactly how alcohol triggers adenosine release is not yet understood. One promising idea is that alcohol disables a molecular pump that normally sucks up free adenosine and transports it back into the cell interior (Gordon et al. 1993).

Alcohol appears to disrupt this adenosine pump, just as it interferes with so many of the brain's other functions. The impairment of this key transporter could leave excess adenosine outside nerve cells, thus explaining the actions mentioned above.

Still, even though the transporter theory seems sound, it is too early to say with confidence that this is, in fact, the long-sought link between alcohol and caffeine (Dunwiddie 1995). Too little is known about how adenosine pumps work and where they are located. Given alcohol's wide-ranging effects, the mechanism behind the observed buildup of adenosine could lie someplace else entirely.

Regardless of *how* alcohol and adenosine are connected, there is little doubt that the connection exists. And that raises an obvious question: If alcohol intoxication involves increased adenosine levels, shouldn't caffeine counteract drunkenness?

Antagonism

Caffeine and alcohol have been used as antidotes for each other for centuries. In the early years of coffee's introduction to Europe, for instance, the French writer Sylvestre Dufour described the following situation in his book *Traitez nouveau et curieus du café, du the, et du chocolat* (1671): "Coffee sobers you up instantaneously, or in any event it sobers up those who are not fully intoxicated. One of my friends who had had too much wine sat down at the gambling table one evening after dinner. He was losing considerable sums, because of having drunk too much wine, he was confusing hearts with diamonds. I took him aside and had him drink a cup of coffee, whereupon he returned to the game with a completely sober head and clear eye." (Schivelbusch, 1992).

We just learned that alcohol apparently raises adenosine levels in the brain, while caffeine *blocks* adenosine receptors and could thus plausibly reverse this effect. Does this mean that neuroscience has verified Dufour's three hundred-year-old observations?

Not exactly.

If adenosine was the only thing that alcohol altered in the brain, then caffeine would, indeed, neatly counteract that action and could be expected to reverse alcohol intoxication. But, as we know, alcohol acts on many more brain systems than just adenosine. While it is affecting adenosine, alcohol is also making GABA receptors more sensitive, and it's inhibiting glutamate receptors, raising dopamine levels, and exerting a wide range of other complicated effects. Meanwhile, caffeine can *only* antagonize adenosine receptors. In a sense, caffeine is fighting with a single sword, while alcohol comes armed with a dozen weapons all flailing at once.

It is estimated that in general only 10 to 20 percent of alcohol's intoxicating effect can be attributed to increased adenosine levels (Dunwiddie 1995). That means that even if you drank enough caffeine to plug every last adenosine receptor in your brain, you would not be staving off more than one-fifth of alcohol-induced inebriation. This is why one is well advised to heed the popular wisdom that caffeine will not offset the effects of alcohol. But, as with most situations involving these two drugs, every rule has an exception. When the amount of alcohol circulating in the brain is low and the amount of caffeine is high, the antagonism of alcohol by caffeine can be significant. In one study, 200 to 400 milligrams of caffeine reversed poor performance on some measures of driving ability in subjects with blood alcohol levels of .04 percent to .06 percent (Moskowitz and Burns 1981). Caffeine has been shown to reverse alcohol-induced decrements in flying-related mental and motor measures and performance on automobile simula-

ors. Some studies of simple reaction time have also shown that
caffeine erases the negative impact of alcohol (Fudin and Nicastro 1988).

Again, the critical caveat to all these studies is that the caffeine doses were always large relative to the alcohol doses. In
fact, the blood alcohol levels in subjects experiencing a reversal
of alcohol-induced performance decrements were all *below* the
.1 percent level that typically defines intoxication. No study
has found that caffeine reverses the effects of alcohol levels at
.1 percent or above (Fudin and Nicastro 1988). A few studies
have even found that when moderate to high levels of alcohol
are involved, caffeine actually *worsens* performance on a variety
of reaction time and vigilance tests (Osborne and Rogers 1983).
Likewise, studies of cognitive performance have shown that on
some types of tests, caffeine *increased* the deleterious effects
of alcohol (Dews 1984).

The general scientific consensus, therefore, is that alcohol
and caffeine interact in complex ways that involve both antagonism and synergism, depending on the dose of both drugs.
Caffeine most clearly offsets the disabling effects of alcohol
when the levels of caffeine are high (above 200 milligrams) and
the levels of alcohol relatively low (below .1% blood alcohol
level). But even under these conditions, the reversal of alcohol's
effects by caffeine is incomplete. Although some of alcohol's
effects are counteracted by caffeine, others remain untouched.
A number of researchers have pointed out the potential danger
of this situation. Perhaps the brain's sleep-regulating center—
an area rich in adenosine receptors—is one place where caffeine best antagonizes the effects of alcohol. Caffeine could,
therefore, make an intoxicated person feel more alert even
though other parts of the brain are considerably impaired. Driving a car or operating dangerous machinery under these conditions would obviously be both irresponsible and hazardous.

The bottom line is that Dufour's centuries-old observatio is both right and wrong. He noted, correctly, that caffeine most useful for those "not fully intoxicated." But his compe ling description of his drunk friend's miraculous recovery to "completely sober head and clear eye" after a single cup of coffee was clearly a case of wishful thinking.

Engineering

We've now seen that alcohol and caffeine do not act in isol; tion from each other at a molecular and neuronal level. I affecting the same key brain neurotransmitter, they are, in fac closely related. This synergy is reflected at the behavioral lev as well.

People often use alcohol and caffeine as complementar tools for mood engineering. To paraphrase a famous advertisin slogan for DuPont, they are the chemicals most often used t achieve "better living." The quote by David Letterman at th beginning of this chapter illustrates the point. Letterma openly and self-consciously consumes a lot of coffee, in par he says, to induce the slightly manic comedic state for whic he is famous (Zehme 1994). Among other professionals wh use caffeine—and alcohol—to prime their creative pumps, pe haps the most widely known are writers, many of whom hav provided eloquent testimony on this practice. In A *Moveab Feast*, for instance, Ernest Hemingway recalls a typical day of writing in 1920s Paris:

> It was a pleasant café, warm and clean and friendly, and I hung up my old waterproof on the coat rack to dry and put my worn and weathered felt hat on the rack above the bench and ordered a cafe au lait. The waiter brought it and I took out a notebook from the pocket of the coat and a pencil and started to write. I

was writing about up in Michigan and since it was a wild, cold, blowing day it was that sort of day in the story ... in the story the boys were drinking and this made me thirsty and I ordered a rum St. James. This tasted wonderful on the cold day and I kept on writing, feeling very well and feeling the good Martinique rum warm me all through my body and my spirit.

Hemingway, despite his consumption of the rum St. James, was proud of his ability to separate his legendary drinking from his writing (Dardis 1989). He generally drank only coffee while he wrote and waited until his notebook was closed for the day before indulging in alcohol.

John Steinbeck, too, usually confined himself to caffeine when writing and relaxed with alcohol. In the journal he kept while writing *The Grapes of Wrath*, he made frequent note of the deleterious effect alcohol had on his work (Demott 1989). "Last night up to Rays' and drank a great deal of champagne," he noted before turning his attention to his unfinished manuscript on June 13, 1938. "I pulled my punches pretty well, but I am not in the dead sober state I could wish." In the next day's entry he says: "Yesterday was a bust. I could have forced the work out but I'd lost the flow of the book and it would have been a weak spot."

F. Scott Fitzgerald wrote his most acclaimed books while downing large helpings of caffeine—usually in the form of cola sodas and coffee. In his later years, he turned to alcohol in increasingly desperate attempts to regain the muse. He began his days drinking pots of coffee, and would then switch to bottles of gin in the afternoon. The resulting pharmacological gyrations didn't help: none of his later works is regarded as equal to *The Great Gatsby* and his other early novels and short stories.

Hemingway, Steinbeck, and Fitzgerald, of course, were alcoholics, and thus drank in far greater quantities and with

much greater intensity than most people. But their musings on
the importance of alcohol and caffeine in their lives have res-
onance for many similarly inclined people. The urge to tinker
with one's mood and energy level by using these yin–yang drugs
is common. Common also is the experience of finding that
caffeine simply exacerbates anxiety, or that alcohol can interfere with emotional intimacy as well as foster it.

To the extent that knowledge can influence behavior, the
information presented in this book about how alcohol and caffeine work might help people use these substances more effectively and intelligently. But would Hemingway have drunk
more moderately had he known how alcohol was affecting his
NMDA receptors? Would Fitzgerald have tempered his cola
consumption if he had known that the caffeine was blocking
adenosine receptors in his brain? It's hard to imagine affirma-
tive answers to these questions. When it comes to drugs—even
one as mild as caffeine—logic and reason can be impressively
useless.

The Multitudes Within

In "Song of Myself," Walt Whitman wrote,

> Do I contradict myself?
> Very well. I contradict myself.
> I am large. I contain multitudes.

Whitman anticipated by more than a hundred years a perspective on human nature that illuminates the often perplexing
relationship people have with their drugs of choice.

The human brain is now known to be a layered and multi-
faceted organ. It is subdivided into discrete functional units
that operate with a great deal of independence. The brain, and

the mind generated by the brain, have been likened to a "society" of more or less autonomous parts (Minsky 1986). Viewed from this perspective, it is not surprising that when it comes to drugs, humans are capable of pronounced contradictions.

The science of the mind is far less developed than the science of the brain, and thus statements about how specific behaviors or cravings emerge from the workings of neurons are necessarily quite tentative. But many neuroscientists have speculated along the following lines. The neocortex—the most recent addition to the human brain from an evolutionary point of view—is the seat of language, music, abstraction, reason, foresight, and reflection. It is speculated that humans use their neocortex to form their sense of who they are—their self-awareness and their self-consciousness. The neocortex "understands" information presented verbally, logically, and sequentially: information such as that presented in books about the nature of alcohol and caffeine, for instance. Other parts of the brain, however, do not work in this way. The limbic system, for example, is believed to support emotions such as empathy, anger, territoriality, aggression, and maternal bonding. And there are brain structures that generate sexual desire, thirst, hunger, pain, pleasure, and other primal sensations. All these structures are the neurological substrates of Whitman's "multitudes."

The conflict generated by the simultaneous activity of all the members of the mind's "society" is, of course, the foundation of much literature and art. It is our capacity for internal conflict and irrationality that defines us as human beings. Fictional characters such as Data, the emotionless android of the television show *Star Trek: The Next Generation*, are compelling precisely because their perfect logic and lack of emotion contrast so sharply with the very imperfect logic of the humans around them.

The point is that drugs such as alcohol and caffeine affect the brain and mind at *all* levels. As we've seen repeatedly, alcohol and caffeine go to work, either directly or indirectly, on the neurotransmitters used in the neocortex, the limbic system, and the dopamine-fueled reward centers, evoking very powerful cravings and sensations that can collide with, or completely overwhelm, more prudent desires generated elsewhere in the brain. An alcoholic reaching yet again for a bottle despite the knowledge that further drinking will be disastrous is responding not to reason or logic, but to deeper voices entirely. Likewise, many coffee drinkers have experienced the tug to have another cup even though they know from past experience that yielding to the temptation is something they're likely to regret.

None of this means that increased knowledge is irrelevant to one's efforts to use alcohol or caffeine wisely. The fact remains that most people are *not* addicted to alcohol or caffeine and can control their consumption to one degree or another. They are neither completely captive to their cravings nor so in control that they don't occasionally drink more alcohol or caffeine than they know is healthy or productive. This suggests that information about how alcohol and caffeine work will be useful to varying degrees for different individuals. A more complete understanding of these substances may help people find ways to use them more effectively in their daily lives. Modern neuroscience suggests, however, that it would be a mistake to discount the multiplicity of the mind, to forget that one's conscious self is not one's entire self, and to ignore the power of the nonrational forces within us.

This deep dichotomy between reason and irrationality can be seen in the world's tremendous appetite for alcohol and caffeine. Alcohol is the liberator of the irrational. Caffeine is the stimulator of the rational. It would appear that the human spirit craves *both* poles and turns to these most familiar of drugs to achieve those ends.

Postscript

Having now written an entire book about alcohol and caffeine, and having mused about their utility for the enhancement of daily living, I feel that I can't ignore an obvious topic: the role these two substances played in the creation of this book and, conversely, the effect the book has had on my use of these two substances.

Most of *Buzz* has been written on caffeine. As I write this sentence, it's 11:12 P.M. I would be asleep were it not for the caffeine molecules coursing through my system. I would, in fact, *prefer* to be asleep. But circumstances have forced me to pursue this book in the wee hours of my life. Thus caffeine, typically administered in a shot or two of evening espresso, has been an invaluable tool. As an experiment, I have tried writing *without* caffeine at times like this, and the results are not pretty. The writing comes out just as tired and flabby as I feel physically.

In contrast, alcohol has played only an indirect role in this book. My ability to write is the first thing to dissolve in alcohol's solvent. If I have as little as a half-glass of wine an hour before writing some critical pressure is lost. With alcohol in my system, I cannot, as Hemingway once said, "close it like the diaphragm of a camera and intensify it so it could be concentrated to the point where the heat shone bright and the smoke began to rise."

I am not, however, a teetotaler. Often, after writing, I indulge in a dram of my favorite scotch or a small shot of good bourbon. And although I forgo wine on nights when I write, I greatly enjoy the gift of Bacchus when mated with the right food. Alcohol, in other words, is a normal and enjoyable part of my life and thus probably deserves mention as playing some kind of supporting role in the book's creation.

But the book has affected my use of alcohol and caffeine just as much as my use of alcohol and caffeine has affected the book. In the past, I sometimes reached for a cup of coffee or a glass of wine as much from habit as from a conscious desire to alter my consciousness. Beverages containing alcohol and caffeine are so embedded in modern society that it is easy to forget that they contain relatively powerful drugs. Now, of course, I can't ignore this fact, and this has made me a more conscious consumer.

In general, I drink less alcohol now than I did prior to beginning this project. I don't automatically take a glass of wine proffered at a party, or assume that if I meet friends at a bar I need to have a beer. In short, I try, with varying degrees of access, to use alcohol deliberately—to enjoy it when the occasion warrants it, and to avoid it when I want a sharper, clearer state of mind.

As for caffeine, I continue my long-standing experiments to find an optimal dose. I love the flavors, the aromas, and the rituals surrounding good coffee and espresso, but I'm also aware that a caffeine buzz is useful for some tasks and not for others. When patience and calm are required (such as when caring for small children), I have found caffeine to be of dubious utility. When the job is clear-cut, or when a slightly manic frame of mind is enjoyable, caffeine can be just the ticket. To remind myself of the ways caffeine affects my mood and personality, I take occasional caffeine holidays of at least a week or two. Knowing how caffeine works in the brain has allowed me to tailor my intake to minimize unpleasant withdrawal symptoms.

In short, two years of research and writing about alcohol and caffeine hasn't convinced me to stop consuming either drug. If I were to boil down the contents of this book to a few intensely-flavored drops of advice, I believe I'd end up with something very similar to some bits of wisdom carved more than 2000 years ago into the stone face of the temple of Apollo at Delphi. Two simple phrases were etched so deeply that to this day you can still read them easily: "know thyself," and "nothing to excess." If the current scientific understanding of alcohol and caffeine says anything, it is exactly that.

References and Suggested Reading

Amerine, M. A., and E. B. Roessler. 1983. *Wines: Their Sensory Evaluation*. New York: Freeman.

Amit, Z., E. A. Sutherland, K. Gill, and S. O. Ogren. 1984. Zimelidine: A review of its effects on alcohol consumption. *Neuroscience and Biobehavioral Reviews* 8:35–54.

Arnaud, M. J. 1993. Metabolism of caffeine and other components of coffee. In *Caffeine, Coffee, and Health*, edited by S. Garattini. New York: Raven.

Ashton, H. 1992. *Brain Function and Psychotropic Drugs*. New York: Oxford University Press.

Athanasiou, R., P. Shaver, and C. Tavris. 1970. Sex: A *Psychology Today* report on more than 20,000 responses to 101 questions about sexual attitudes and practices. *Psychology Today*, February:39–52.

Barr, H. M., and A. P. Streissguth. 1991. Caffeine use during pregnancy and child outcome: A 7–year prospective study. *Neurotoxicology and Teratology* 13:441–448.

Barrett-Connor, E., J. C. Chang, and S. L. Edelstein. 1994. Coffee-associated osteoporosis offset by daily milk consumption. *Journal of the American Medical Association* 271:280–283.

Battig, K., and H. Wetzl. 1993. Psychopharmacological profile of caffeine. In *Caffeine, Coffee, and Health,* edited by S. Garattini. New York: Raven.

Benowitz, N. L., S. M. Hall, and G. Modin. 1989. Persistent increase in caffeine concentrations in people who stop smoking. *British Medical Journal* 298:1075–1076.

Berry, H., Beverage Marketing Corporation, New York 1994. Personal communications.

Birnbaum, I. M., E. S. Parker, and J. T. Hartley. 1978. Alcohol and human memory: Retrieval processes. *Journal of Verbal Learning and Verbal Behavior* 17:325–335.

Blitzer, R. D. 1994. Personal communications.

Blitzer, R. D., O. Gil, and E. M. Landau. 1990. Long-term potentiation in rat hippocampus is inhibited by low concentrations of alcohol. *Brain Research* 537:203–208.

Burke, E. R. 1992. *Cycling Health and Physiology.* Brattleboro, Vt.: Vitesse Press.

Caffeine as a downer: Rare reaction to caffeine. 1990. *Harvard Medical School Health Letter,* April:7.

Charness, M. E. 1994. Personal communications.

Charness, M. E., R. M. Safran, and G. Perides. 1994. Ethanol inhibits neural cell–cell adhesion. *Journal of Biological Chemistry* 269:9304–9309.

Charness, M. E., R. P. Simon, and D. A. Greenberg. 1989. Alcohol and the nervous system. *New England Journal of Medicine* 321:442–454.

Chen, C.-C., and E.-K. Yeh. 1989. Population differences in ALDH levels and flushing response. In *Molecular Mechanisms of Alcohol,* edited by G. Y. Sun. New York: Humana.

Chou, T. 1992. Wake up and smell the coffee: Caffeine, coffee, and the medical consequences. *Western Journal of Medicine* 157:544–554.

Cicero, T. J. 1994. Effects of paternal exposure to alcohol on offspring development. *Alcohol Health and Research World* 18:37–41.

Cloninger, C. R., M. Bohman, and S. Siguardsson. 1981. Inheritance of alcohol abuse: Cross-fostering analysis of adopted men. *Archives of General Psychiatry* 38:861–868.

Cohen, S., D. A. Tyrrell, M. A. Russell, J. J. Jarvis, and A. P. Smith. 1993. Smoking, alcohol consumption, and susceptibility to the common cold. *American Journal of Public Health* 83:1277–1283.

Coles, C. 1994. Critical periods for prenatal alcohol exposure. *Alcohol Health and Research World* 18:22–29.

Costill, D. L., G. Dalasky, and W. Fink. 1978. Effects of caffeine ingestion on metabolism and exercise performance. *Medicine and Science in Sports and Exercise* 10:155–158.

Crabbe, J. C., J. K. Belknap, and K. J. Buck. 1994. Genetic animal models of alcohol and drug abuse. *Science* 264:1715–1723.

Daly, J. W. 1993. Mechanism of action of caffeine. In *Caffeine, Coffee, and Health,* edited by S. Garattini. New York: Raven.

Daly, J. W., D. Shi, V. Wong, and O. Nikodijevic. 1994. Chronic effects of ethanol on central adenosine function of mice. *Brain Research* 650:153–156.

Dardis, T. 1989. *The Thirsty Muse.* New York: Ticknor & Fields.

Demott, R. 1989. *Working Days: The Journals of "The Grapes of Wrath."* New York: Penguin.

Department of Health and Human Services, National Institute on Alcohol Abuse and Alcoholism. 1993. *Alcohol and Health: Eighth Special Report to the U. S. Congress.*

Dews, P. B. 1984. *Caffeine: Perspectives from Recent Research.* New York: Springer-Verlag.

Di Chiara, G., and A. Imperato. 1988. Drugs abused by humans preferentially increase synaptic dopamine concentrations in the mesolimbic system of freely moving rats. *Proceedings of the National Academy of Sciences* 85:5274–5278.

Dolnick, E. 1990. Le paradox français. *Health,* May–June:41–47.

Dowling, J. E. 1992. *Neurons and Networks.* Cambridge, Mass.: Harvard University Press.

Dumas, M. 1982. Systematic determination of caffeine plasma concentrations at birth in pre-term and full-term infants. *Developmental Pharmacology and Therapeutics* 4:182–186.

Dunwiddie, T. 1995. Personal communications.

Eskenazi, B. 1993. Caffeine during pregnancy: Grounds for concern? *Journal of the American Medical Association* 270:2973–2974.

Evans, S. M., and R. R. Griffiths. 1992. Caffeine tolerance and choice in humans. *Psychopharmacology* 108:51–59.

Fastbom, J., and B. B. Fredholm. 1990. Effects of long-term theophylline treatment on adenosine A_1 receptors in rat brain: Autoradiographic evidence for increased receptor number and altered coupling to G-proteins. *Brain Research* 507:195–199.

Fredholm, B. B. 1984. Effects of methylxanthines on skeletal muscle and on respiration. In *The Methylxanthine Beverages and Foods: Chemistry, Consumption, and Health Effects,* edited by G. A. Spiller. New York: Liss.

Frezza M., C. DiPadova, G. Pozzato, M. Terpin, E. Baraona, and C. S. Lieber. 1990. High blood alcohol levels in women: The role of decreased gastric alcohol dehydrogenase activity and first-pass metabolism. *New England Journal of Medicine* 322:95–99.

Friedman, G. D., and A. L. Klatsky. 1993. Is alcohol good for your health? *New England Journal of Medicine* 329:1882–1883.

Friedman, H. J., J. A. Carpenter, D. Lester, and C. L. Randall. 1980. Effect of alpha-methyl-p-tyrosine on dose-dependent mouse strain differences in locomotor activity after alcohol. *Journal of Studies on Alcohol* 41:1–7.

Frye, G. D., and G. R. Breese. 1981. An evaluation of the locomotor stimulating action of alcohol in rats and mice. *Psychopharmacology* 75:372–379.

Fudin, R., and R. Nicastro. 1988. Can caffeine antagonize alcohol-induced performance decrements in humans? *Perceptual and Motor Skills* 67:375–391

Galletly, D. C., M. Fennelly, and J. G. Whitwam. 1989. Does caffeine withdrawal contribute to postanaesthetic morbidity? *Lancet* 1:1335.

Garcia, R. 1993. The cardiovascular effects of caffeine. In *Caffeine, Coffee, and Health*, edited by S. Garattini. New York: Raven.

Gianoulakis, C., P. Angelogianni, M. Meany, J. Thavundayil, and V. Tawar. 1990. Endorphins in individuals with high and low risk for development of alcoholism. In *Opioids, Bulimia, and Alcohol Abuse and Alcoholism*, edited by L. D. Reid. New York: Springer-Verlag.

Gilbert, R. M. 1984. Caffeine consumption. In *The Methylxanthine Beverages and Foods: Chemistry, Consumption, and Health Effects*, edited by G. A. Spiller. New York: Liss.

Gilbert, R. M. 1992. *Caffeine, the Most Popular Stimulant*. New York: Chelsea House.

Gittelman, R. 1983. Experimental and clinical studies of stimulant use in hyperactive children and children with other behavioral disorders. In *Stimulants: Neurochemical, Behavioral and Clinical Perspectives*, edited by I. Creese. New York: Raven.

Glass, R. M. 1994. Caffeine dependence: What are the implications? *Journal of the American Medical Association* 272:1065–1066.

Goldstein, A. 1994. *Addiction, from Biology to Drug Policy*. New York: Freeman.

Goodwin, D. W. 1985. Alcoholism and genetics: The sins of the fathers. *Archives of General Psychiatry* 42:171–174.

Gordis, E., M. C. Dufour, K. R. Warren, R. J. Jackson, R. L. Floyd, and D. W. Hungerford. 1995. Should physicians counsel patients to drink alcohol? *Journal of the American Medical Association* 273:1415.

Gordis, L. 1990. Consumption of methylxanthine-containing beverages and risk of pancreatic cancer. *Cancer Letters* 52:1–12.

Gordon, A., M. K. Sapru, S. W. Krauss, and I. Diamond. 1993. Nucleoside transport and ethanol-induced heterologous desensitization. In *Alcohol, Cell Membranes, and Signal Transduction in Brain*, edited by C. Alling et al. New York: Plenum.

Graham, H. N. 1984. Tea: The plant and its manufacture: Chemistry and consumption of the beverage. In *The Methylxanthine Beverages and Foods: Chemistry, Consumption, and Health Effects*, edited by G. A. Spiller. New York: Liss.

Grobbee, D. E., E. B. Rimm, E. Giovannucci, G. Colditz, M. Stampfer, and W. Willett. 1990. Coffee, caffeine, and cardiovascular disease in men. *New England Journal of Medicine* 323:1026–1032.

Grounds for breaking the coffee habit? 1990. *Tufts University Diet & Nutrition Letter*, February:3.

Harburg, E., D. R. Davis, and R. Caplan. 1982. Parent and offspring alcohol use:

Imitative and aversive transmission. *Journal of Studies on Alcohol* 43:497–516.

Harper C. G., and J. J. Krill. 1990. Neuropathology of alcoholism. *Alcohol and Alcoholism* 25:207–216.

Higuci, S., S. Matsushita, H. Imazeki, T. Kinoshita, S. Takagi, and H. Kono. 1994. Aldehyde dehydrogenase genotypes in Japanese alcoholics. *Lancet* 343:741–742.

Holden, C. 1994. A cautionary genetic tale: The sobering story of D_2. *Science* 264:1696.

Hollister, L. E. 1975. The mystique of social drugs and sex. In *Sexual Behavior, Pharmacology, and Biochemistry*, edited by M. Sandler and G. L. Gessa. New York: Raven.

Hooper, J., and D. Teresi. 1987. *The Three-Pound Universe.* New York: Dell.

Horgan, J. 1993. Eugenics revisited. *Scientific American*, June:123–131.

Hughes, J. R. 1992. Clinical importance of caffeine withdrawal. *New England Journal of Medicine* 327:1160–1161.

Infante-Rivard C., A. Fernandez, R. Gauthier, M. David, and G. E. Rivard. 1993. Fetal loss associated with caffeine intake before and during pregnancy. *Journal of the American Medical Association* 270:2940–2943.

Iversen, S. D., and L. L. Iversen. 1981. *Behavioural Pharmacology.* New York: Oxford University Press.

Jacobson, J. L., and S. W. Jacobson. 1994. Prenatal alcohol exposure and neurobehavioral development. *Alcohol Health and Research World* 18:30–36.

Jacobson S. W., G. G. Fein, J. L. Jacobson, P. M. Schwartz, and J. K. Dowler. 1984. Neonatal correlates of prenatal exposure to smoking, caffeine, and alcohol. *Infant Behavioral Development* 7:253–265.

James, J. 1991. *Caffeine and Health.* New York: Academic Press.

Jensen, G. B., and B. Pakkenberg. 1993. Do alcoholics drink their neurons away? *Lancet* 342:1201–1204.

Johnson, G. 1991. *In the Palaces of Memory.* New York: Knopf.

Jones, B. M. 1973. Memory impairment on the ascending and descending limbs of the blood alcohol curve. *Journal of Abnormal Psychology* 82:24–32.

Josselyn, S. A., and R. J. Beninger. 1991. Behavioral effects of intrastriatal caffeine mediated by adenosinergic modulation of dopamine. *Pharmacology, Biochemistry and Behavior* 39:97–103.

Kaplan, G. 1995. Personal communications.

Kaplan, G., D. J. Greenblatt, M. A. Kent, and M. M. Cotreau-Bibbo. 1993. Caffeine treatment and withdrawal in mice: Relationships between dosage, concentrations, locomotor activity and A_1 adenosine receptor binding. *Journal of Pharmacology and Experimental Therapeutics* 266:1563–1572.

Karni, A., D. Tanne, B. S. Rubenstein, J. J. M. Askenasy, and D. Sagi. 1994. Dependence on REM sleep of overnight improvement of a perceptual skill. *Science* 265:679–682.

Kendler K., A. C. Heath, M. C. Neale, R. C. Kessler, and L. J Eaves. 1992. A population-based twin study of alcoholism in women. *Journal of the American Medical Association* 268:1877–1882.

Kihlman, B. A. 1977. *Caffeine and Chromosomes*. Oxford: Elsevier.

Koob, G. R., and F. E. Bloom. 1988. Cellular and molecular mechanisms of drug dependence. *Science* 242:715–719.

Kurppa, K, P. C. Holmberg, E. Kuosma, and L. Saxen. 1983. Coffee consumption during pregnancy and selected congenital malformation: A nationwide case-control study. *American Journal of Public Health* 73:1397–1399.

Lang, A. R., D. J. Goeckner, V. J. Adesso, and G. A. Marlatt. 1975. Effects of alcohol on aggression in male social drinkers. *Journal of Abnormal Psychology* 84:418–425.

Lang, A. R., J. Searles, R. Lauerman, and V. Adesso. 1980. Expectancy, alcohol, and sex guilt as determinants of interest in and reaction to sexual stimuli. *Journal of Abnormal Psychology* 89:644–653.

Laska E., A. Sunshine, F. Mueller, W. B. Elvers, C. Siegel, and A. Rubin. 1984. Caffeine as an analgesic adjuvant. *Journal of the American Medical Association* 251:1711–1718.

Lawrin, M. O., C. A. Naranjo, and E. M. Sellers. 1986. Identification of new drugs for modulating alcohol consumption. *Psycho-pharmacology Bulletin* 22: 1020–1025.

Lehman, B. A. 1989. A tempest in a coffee pot. *Boston Globe*, November 27:25.

Leibel, R. L., M. Rosenbaum, and J. Hirsch. 1995. Changes in energy expenditure resulting from altered body weight. *New England Journal of Medicine* 332: 621–628.

Lewin, L. 1931 [1964]. *Phantastica: Narcotics and Stimulating Drugs*. New York: Dutton.

Livermore, B. 1991. Caffeine boosts eating disorders. *Health*, June:16.

Lovinger, D. M., and R. W. Peoples. 1993. Actions of alcohols and other sedative/hypnotic compounds on cation channels associated with glutamate and 5–HT3 receptors. In *Alcohol, Cell Membranes, and Signal Transduction in Brain*, edited by C. Alling et al. New York: Plenum.

Makowsky, C. R., and P. C. Whitehead. 1991. Advertising and alcohol sales: A legal impact study. *Journal of Studies on Alcohol* 52:555–567.

Mann, C. 1994. Behavioral genetics in transition. *Science* 264:1686–1689.

Masters, W. H., and V. E. Johnson. 1986. *Masters and Johnson on Sex and Human Loving*. Boston: Little, Brown.

McClearn, G. E., and D. A. Rodgers. 1959. Differences in alcohol preference among inbred strains of mice. *Journal of Studies on Alcohol* 20:691–695.

McCoy, E., and J. F. Walker. 1991. *Coffee and Tea*. New York: Theron Raines.

Meadows, G. G., S. E. Blank, and D. D. Duncan. 1989. Influence of alcohol consumption on natural killer cell activity in mice. *Alcoholism: Clinical and Experimental Research* 13:476–479.

Miller, M. 1994. Call for a daily dose of wine ferments critics. *Wall Street Journal*, June 17:B1.

Mills J. L., L. B. Holmes, J. H. Aarons, et al. 1993. Moderate caffeine use and the risk of spontaneous abortion and intrauterine growth retardation. *Journal of the American Medical Association* 269:593–597.

Minsky, M. 1986. *The Society of Mind*. New York: Simon and Schuster.

skowitz, H., and M. Burns. 1981. The effects of alcohol and caffeine, alone and in combination, on skills performance. In *Alcohol, Drugs and Traffic Safety*, edited by L. Goldberg. Stockholm: Almqvist & Wiksel.

urphy, J. M., W. J. McBride, L. Lumeng, and T. -K. Li. 1987. Contents of monoamines in forebrain regions of alcohol-preferring and non-preferring lines of rats. *Pharmacology, Biochemistry and Behavior* 26:389–392.

ranjo, C. A., K. E. Kadlec, P. Sanhueza, D. Woodley-Remus, and E. M. Sellers. 1990. Fluoxetine differentially alters alcohol intake and other consummatory behaviors in problem drinkers. *Clinical Pharmacology and Therapeutics* 47:490–498.

sh, H. 1962. *Alcohol and Caffeine: A Study of Their Psychological Effects*. Springfield, Ill.: Thomas.

thanson, J. 1984. Caffeine and related methylxanthines: Possible naturally occuring pesticides. *Science* 226:184–187.

ational Coffee Association. 1991. Coffee-drinking study.

ehlig, A., J. L. Daval, and G. Debry. 1992. Caffeine and the central nervous system: Mechanisms of action, biochemical, metabolic and psychostimulant effects. *Brain Research Reviews* 17:139–170.

IH consensus development panel on optimal calcium intake. 1994. *Journal of the American Medical Association* 272:1942–1948.

oble, E. P., K. Blum, R. Ritchie, A. Montgomery, and P. J. Sheridan. 1991. Allelic association of the D_2 dopamine receptor gene with receptor binding characteristics in alcoholism. *Archives of General Psychiatry* 48:648–654.

sborne, D. J., and Y. Rogers. 1983. Interactions of alcohol and caffeine on human reaction time. *Aviation, Space and Environmental Medicine* 54:528–534.

lca, J. 1989. Sleep researchers awake to possibilities. *Science* 245:351.

rsons, W. D., and A. H. Neims. 1978. Effect of smoking on caffeine clearance. *Clinical Pharmacology and Therapeutics* 24:40–45.

illis, John W. 1989. Caffeine and premenstrual syndrome. *American Journal of Public Health* 79:1680.

ckens, R. W., D. S. Svikis, M. McGue, D. T. Lykken, L. L. Hesten, and P. J. Clayton. 1991. Heterogeneity in the inheritance of alcoholism. *Archives of General Psychiatry* 48:19–28.

octor, W. R., and T. V. Dunwiddie. 1984. Behavioral sensitivity to purinergic drugs parallels ethanol sensitivity in selectively bred mice. *Science* 224:519–521.

ainnie, D. G., H. Grunze, R. W. McCarley, and R. W. Greene. 1994. Adenosine inhibition of mesopontine cholinergic neurons: Implications for EEG arousal. *Science* 263:689–692.

egestein, Q. 1995. Personal communications.

euters News Service. 1994. Germany's Gerasch banned for two years. January 26.

ichards, B. 1995. Cafe au revoir? Some say coffee has become too cool. *Wall Street Journal*, January 31:A1.

isto R., R. T. Gentry, R. Hernandez-Munoz, E. Baraona, and C. S. Lieber. 1990. Aspirin increases blood alcohol concentrations in humans after ingestion of ethanol. *Journal of the American Medical Association* 264:2406–2408.

Robson, R. A. 1992. The effects of quinolines on xanthine pharmacokinetics. *Ame*
ican *Journal of Medicine* 92:22.

Rossignol, A. M., H. Bonnlander, L. Song, and J. W. Phillis. 1991. Do women wit
premenstrual symptoms self-medicate with caffeine? *Epidemiology* 2:403
408.

Salvatore, C. A. 1993. Molecular cloning and characterization of the human *A*
adenosine receptor. *Proceedings of the National Academy of Sciences.* 9ℓ
10365–10369.

Sanders, R. C., and T. R. Murray. 1988. Chronic theophylline exposure increase
agonist and antagonist binding to A_1 adenosine receptors in rat brain. *Neι*
ropharmacology 27:757–760.

Saxena, Q., R. Saxena, and W. Adler. 1981. Regulation of natural killer activity *i*
vivo: High natural killer activity in alcohol-drinking mice. *Indian Journal ι*
Experimental Biology 19:1001–1006.

Schivelbusch, W. 1992. *Tastes of Paradise: A Social History of Spices, Stimulant*
and Intoxicants. New York: Pantheon.

Schuckit, M. A. 1985. The clinical implications of primary diagnostic groups amon
alcoholics. *Archives of General Psychiatry* 42:1043–1049.

Seale T. W., K. A. Abla, M. T. Shamim, J. M. Carney, and J. W. Daly. 1988. 3,
dimethyl-1-propargylxanthine: A potent and selective *in vivo* antagonist ι
adenosine analogs. *Life Sciences* 43:1671–1684.

Seitz, H., G. Egerer, and U. Simanowski. 1990. High blood alcohol levels in womeι
New England Journal of Medicine 323:58.

Shi, D., O. Nikodijevic, K. A. Jacobson, and J. W. Daly. 1993. Chronic caffeiι
alters the density of adenosine, adrenergic, cholinergic, GABA, and seroteι
nin receptors and calcium channels in mouse brain. *Cellular and Moleculι*
Neurobiology 13:247–261.

Siegel, R. K. 1989. *Intoxication: Life in Pursuit of Artificial Paradise.* New Yorι
Dutton.

Silverman, K., S. M. Evans, E. C. Strain, and R. R. Griffiths. 1992. Withdrawι
syndrome after the double-blind cessation of caffeine consumption. *Ne*
England Journal of Medicine 327:1109–1114.

Smart, R. G. 1988. Does alcohol advertising affect overall consumption? A revieι
of empirical studies. *Journal of Studies on Alcohol* 49:314–323.

Snel, J. 1993. Sleep and wakefulness. In *Caffeine, Coffee, and Health*, edited by ℓ
Garattini. New York: Raven.

Stradling, J. R. 1993. Recreational drugs and sleep. *British Medical Journι*
306:573.

Strain, E. C., G. K. Mumford, K. Silverman, and R. R. Griffiths. 1994. Caffeiι
dependence syndrome. *Journal of the American Medical Associatic*
272:1043–1048.

Takeshita, T., K. Morimoto, X.-Q. Mao, T. Hashimoto, and J. Furyuama. 199ᦆ
Phenotypic differences in low K_m aldehyde dehydrogenase in Japanese worι
ers. *Lancet* 341:837–838.

Tanaka, Y. 1990. Effect of xanthine derivatives on hippocampal long-term poteι
tiation. *Brain Research* 522:63–68.

Tice, P. M. 1992. *Altered States: Alcohol and Other Drugs in America*. Rochester, N.Y.: Strong Museum.

Treistman, S. N., M. M. Moynihan, and D. E. Wolf. 1987. Influence of alcohols, temperature, and region on the mobility of lipids in neuronal membrane. *Biochimica et Biophysica Acta* 898:109–120.

Uhl, G., K. Blum, E. Noble, and S. Smith. 1993. Substance abuse vulnerability and D_2 receptor genes. *Trends in Neurosciences* 16:83–88.

Viani, R. 1993. The composition of coffee. In *Caffeine, Coffee, and Health*, edited by S. Garattini. New York: Raven.

Weight, F. F. 1994. Personal communications.

Weight, F. F. 1992. Cellular and molecular physiology of alcohol actions in the nervous system. *International Review of Neurobiology* 33:289–348.

Weight, F. F., R. W. Peoples, J. M. Wright, C. Li, L. G. Aguaya, D. M. Lovinger, and G. White. 1993. Neurotransmitter-gated ion channels as molecular sites of alcohol action. In *Alcohol, Cell Membranes, and Signal Transduction in Brain*, edited by C. Alling et al. New York: Plenum.

Weir, J., T. D. Noakes, K. Myburgh, and B. Adams. 1987. A high carbohydrate diet negates the metabolic effects of caffeine during exercise. *Medicine and Science in Sports and Exercise* 19:100–105.

Wilford, J. N. 1992. Jar in Iranian ruins betrays beer drinkers of 3500 B.C. *New York Times*, February 13:A16.

Wilson, G. T., and D. M. Lawson. 1976. Expectancies, alcohol, and sexual arousal in male social drinkers. *Journal of Abnormal Psychology* 85:587–594.

Yesair, D. W. 1984. Human disposition and some biochemical aspects of methylxanthines. In *The Methylxanthine Beverages and Foods: Chemistry, Consumption, and Health Effects*, edited by G. A. Spiller. New York: Liss.

Zehme, Bill. 1994. Letterman lets his guard down. *Esquire*, December:97–102.

Zielke, C. L., and Zielke, H. R. 1987. Chronic exposure to subcutaneously implanted methylxanthines. *Biochemical Pharmacology* 36:2533–2538.